An Outline of Strategies for Building an Innovation System for Knowledge City

Keith Crane, Howard J. Shatz, Shanthi Nataraj,
Steven W. Popper, Xiao Wang

Sponsored by the Guangzhou Development District

RAND Environment, Energy, and Economic Development
A RAND INFRASTRUCTURE, SAFETY, AND ENVIRONMENT PROGRAM

This project was sponsored by the Guangzhou Development District and was conducted in the Environment, Energy, and Economic Development Program within RAND Infrastructure, Safety and Environment.

Library of Congress Control Number: 2012949157

ISBN: 978-0-8330-7700-4

Published 2012 by the RAND Corporation
1776 Main Street, P.O. Box 2138, Santa Monica, CA 90407-2138
1200 South Hayes Street, Arlington, VA 22202-5050
4570 Fifth Avenue, Suite 600, Pittsburgh, PA 15213-2665
RAND URL: http://www.rand.org/
To order RAND documents or to obtain additional information, contact
Distribution Services: Telephone: (310) 451-7002;
Fax: (310) 451-6915; Email: order@rand.org

Preface

China's Guangzhou Development District (GDD) is focused on creating an environment conducive to innovation in Sino-Singapore Guangzhou Knowledge City, a new project being carried out by GDD with Singbridge of Singapore. Knowledge City is to be a new environmentally and technologically advanced city that hosts innovative industries and their associated knowledge workers.

This document outlines a strategy to help GDD succeed in its efforts. It presents specific actions that GDD should undertake in three broad areas: attracting and retaining high-technology companies; attracting and retaining highly skilled, innovative workers; and ensuring the availability of innovation-oriented financing. It then ranks them by importance, ease of implementation, and timing. The outline provides GDD with a roadmap for working toward the successful establishment of Knowledge City.

This report is a companion volume to another RAND report, *Creating an Innovation System for Knowledge City* (TR-1293-GDD), a compilation of results from the interim analyses conducted for this project and supporting evidence for the conclusions presented in this report. That volume is available at http://www.rand.org/pubs/technical_reports/TR1293.html.

This project was sponsored by GDD. The report should be of interest to GDD and Guangzhou officials who are responsible for the success of Knowledge City, researchers and government officials who focus on innovation-based economic development, and anyone studying or involved in the economic transformation of China.

The RAND Environment, Energy, and Economic Development Program

This project was conducted in the Environment, Energy, and Economic Development Program (EEED) within RAND Infrastructure, Safety and Environment (ISE). The mission of RAND Infrastructure, Safety, and Environment is to improve the development, operation, use, and protection of society's essential physical assets and natural resources and to enhance the related social assets of safety and security of individuals in transit and in their workplaces and communities. The EEED research portfolio addresses environmental quality and regulation, energy resources and systems, water resources and systems, climate, natural hazards and disasters, and economic development—both domestically and internationally. EEED research is conducted for government, foundations, and the private sector.

Questions or comments about this report should be sent to the project leaders, Debra Knopman (Debra_Knopman@rand.org), Keith Crane (Keith_Crane@rand.org), or Howard Shatz (Howard_Shatz@rand.org). Information about the Environment, Energy, and Economic Development Program is available online (http://www.rand.org/ise/environ). Inquiries about EEED projects should be sent to the following address:

Keith Crane, Director
Environment, Energy, and Economic Development Program, ISE
RAND Corporation
1200 South Hayes Street
Arlington, VA 22202-5050
703-413-1100, x5520
Keith_Crane@rand.org

Contents

Figures

Tables

Summary

China's Guangzhou Development District (GDD) is focused on creating an environment conducive to innovation in a new development called Sino-Singapore Guangzhou Knowledge City (see Figure S.1). Jointly developed by GDD and Singbridge of Singapore, Knowledge City is to be a new environmentally and technologically advanced city that hosts innovative industries and their associated knowledge workers. The project is designed to spur the transformation of Guangzhou from less-complex manufacturing to higher-wage research- and innovation-based knowledge industries.

When completed, Knowledge City will be one of several components of GDD, a specially designated economic district in the Luogang District of Guangzhou. GDD includes Guangzhou Science City (also part of the Guangzhou High-Technology Industrial Development Zone), the Guangzhou Economic and Technological Development District, the Guangzhou Free Trade Zone, the Guangzhou Export Processing Zone, and Guangzhou International Biological Island.

This document outlines a strategy for Knowledge City. It stems from research and analysis conducted by the RAND Corporation in collaboration with GDD between May 2011 and May 2012. This report is a companion volume to another RAND report, *Creating an Innovation System for Knowledge City* (Nataraj et al., 2012), a compilation of results from the interim analyses conducted for this project and supporting evidence for the conclusions presented here.

Figure S.1
Proposed Knowledge City

SOURCE: Sino-Singapore Guangzhou Knowledge City.
RAND *MG1240-S.1*

Outline of the Strategy: The Situation, Guiding Policy, and Actions

To achieve its goal of establishing Knowledge City as a new environmentally and technologically advanced city that hosts innovative industries and their associated knowledge workers, GDD must convince advanced companies and talented individuals to stake their futures on an unproven new development. To convince them, GDD must instill confidence in potential partners that it possesses the attributes to drive the success of an innovation-based cluster. Emphasizing the attractiveness of its existing assets is one way to instill such confidence. Another is to create innovation-friendly assets that other regions will find hard to replicate. Creating a reputation for having these attributes could assist in attracting innovation-oriented firms to Knowledge City; the presence of these firms would reinforce Knowledge City's reputation for innovation, thus attracting more innovation-oriented firms to the area and creating a virtuous circle that would make it difficult for other regions to catch up.

The strategy for Knowledge City needs to achieve three goals to foster the creation of an innovative area:

1. Attract high-technology companies and enable their growth.
2. Attract and retain highly skilled, innovative people.
3. Ensure the availability of innovation-oriented financing.

Companies, people, and financing form the three pillars of an innovation system. Underlying them are the legal and regulatory environment and the business support environment. When all five of these elements come together to support innovation, the result is a cluster of innovative high-technology firms.

The actions that we present as foundational to a strategy focus on companies, people, and financing. They also have a direct effect on the presence of innovative companies, talented workers, and innovation-oriented financing and improve both the legal and regulatory environment and the business support environment.

A Roadmap for Strategic Actions

Based on our research, we propose that GDD take action in several policy areas to fulfill each of its three goals. Tables S.1, S.2, and S.3 denote the specific policy area into which each action falls; the priority for implementation (high, medium, or low); the ease of implementation (easy, medium, or difficult); and the order of sequencing (early, middle, or late). In each goal area (companies, people, and financing), we suggest that the actions be taken in the order listed.

Attract High-Technology Companies and Enable Their Growth

GDD will want to market Knowledge City, attract an anchor institution or institutions (defined as a prominent company, research institute, or university that will attract researchers and suppliers), and improve the overall innovation environment. Table S.1 summarizes the actions and policies to support the achievement of this goal.

Table S.1
Summary of Actions to Attract High-Technology Companies and Enable Their Growth

Policy Area	Action	Priority	Ease	Sequencing
Marketing	Hire a marketing or public relations company. GDD will need to make Knowledge City known to prospective investors and workers. Competition for investors and knowledge workers is fierce: Professional assistance will be needed to craft the message.	High	Easy	Early
Anchor institutions	Attract an anchor institution. An anchor institution will attract researchers and suppliers and will increase the business community's confidence in the success of Knowledge City.	High	Difficult	Early
Innovation environment	Set up an office to assist companies with all legal, administrative, and financing issues. Such an office should track all relevant laws and regulations, all available incentive programs, and the activities of competitor regions.	Medium	Medium	Early
Innovation environment	Establish a "case officer" system for each company, eventually housing the case officers in a general assistance office. The case officer can provide personal assistance to each company and help companies obtain the benefits for which they qualify.	Medium	Easy	Early

Table S.1—Continued

Policy Area	Action	Priority	Ease	Sequencing
Innovation environment	Have the office and case officer help companies apply for and receive all tax and non-tax incentives for which they are eligible. The office and case officer can help Knowledge City companies more easily work through government requirements.	Medium	Medium	Early
Innovation environment	Ensure that non-tax incentive packages, such as grants, loan subsidies, or direct purchases by GDD on behalf of firms, can respond to the unique needs of innovative firms. Although assistance with land and buildings can always help, non-tax incentives should also include help with making expensive technical equipment or laboratory space available to firms.	Medium	Easy	Middle
Innovation environment	Support initiatives by high-technology firms in Knowledge City to market their products to East Asia. Selling to demanding markets like those in East Asia will spur product innovation and improvement.	Medium	Medium	Middle
Innovation environment	Make Knowledge City a zone of intellectual property rights (IPR) enforcement. Weak IPR enforcement can affect the quality and type of investment. Ensuring that intellectual property (IP) will be protected can make Knowledge City more desirable to innovators.	Medium	Difficult	Middle

Table S.1—Continued

Policy Area	Action	Priority	Ease	Sequencing
Marketing	Sponsor an annual international trade fair or conference. The purposes of such a trade fair would be to highlight Knowledge City as a location for the particular industry and to create connections between Knowledge City businesses and suppliers and customers from around the world.	Low	Easy	Late
Innovation environment	Expand assistance for international patenting. Having Chinese-origin inventors pursue patents in the most developed countries can help inventors upgrade the quality and originality of their inventions.	Low	Medium	Late

Attract and Retain Innovative People

GDD will want to ensure that the quality of life in Knowledge City is high, making it more attractive to knowledge workers. GDD will also want to adopt policies to directly attract talent and returnees (i.e., local workers returning from overseas), enhance labor flexibility, and facilitate the creation of networks. Table S.2 summarizes the actions and policies to support the achievement of this goal.

Table S.2
Summary of Actions to Attract and Retain Innovative People

Policy Area	Action	Priority	Ease	Sequencing
Workers and networks	Establish an advisory council. Drawn from investors and successful entrepreneurs, such a council should provide GDD with valuable advice and form the heart of a network.	High	Easy	Early

Table S.2—Continued

Policy Area	Action	Priority	Ease	Sequencing
Quality of life	Ensure that rail transit connections to Knowledge City are completed and that trains run frequently. Rapid, high-quality public transport will enable people to live where they want while still conveniently commuting to Knowledge City.	High	Difficult	Early
Workers and networks	Increase efforts to attract returnees to Knowledge City, possibly by hiring a professional recruitment firm. Returnees often have special skills that can boost innovation in Knowledge City.	High	Medium	Early
Quality of life	Attract high-quality middle and elementary schools to Knowledge City. Talented workers will want to send their children to good schools and will want to live where they can find such schools.	High	Medium	Middle
Quality of life	Attract a destination shopping center to Knowledge City and facilitate the creation of entertainment venues. These retail and entertainment establishments will make Knowledge City a more desirable place to live and work and can facilitate the formation of networks as well.	Medium	Difficult	Middle
Workers and networks	Facilitate the development of networks among people who have received assistance from GDD by creating opportunities for them to meet. Networks will allow for more informal information sharing and will help generate ideas.	Medium	Medium	Middle

Table S.2—Continued

Policy Area	Action	Priority	Ease	Sequencing
Workers and networks	Facilitate cooperation with overseas Chinese professionals from Guangzhou and Guangdong Province. Overseas Chinese professionals can share their unique skills and open up channels for investment and trade.	Medium	Medium	Middle
Workers and networks	Work on a local and regional basis to smooth out non-legal aspects of laying off workers and shutting down failing firms. For example, establish a job training and placement program. Labor flexibility will enable employers to find the right workers and workers to find the right jobs, if they have lost their jobs. GDD can enhance flexibility by providing information on employment opportunities and otherwise assisting laid-off workers in finding new jobs.	Low	Medium	Late
Workers and networks	Work with provincial and national governments to improve the legal environment for laying off workers and shutting down failing firms. Businesses in GDD find it difficult to lay off workers, but such labor flexibility is important for innovative areas. Longer-term legal reforms may be necessary.	Low	Difficult	Late
Quality of life	Create events to make Knowledge City an exciting destination. Not only will such events make Knowledge City a more desirable place to live and work, but they could help with marketing.	Low	Easy	Late

Table S.2—Continued

Policy Area	Action	Priority	Ease	Sequencing
Workers and networks	Review and improve non-compete clauses. When workers change jobs, they may bring valuable skills and knowledge to their new companies. Non-compete clauses in labor contracts are important, but if they are too strict they could limit valuable information spillovers. GDD will want to ensure that non-compete clauses protect IP but also foster the dissemination of knowledge and skills.	Low	Difficult	Late

Ensure the Availability of Innovation-Oriented Financing

GDD will want to increase the availability of commercial, innovation-oriented financing with appropriate regard to risk and ensure the presence of competitive commercial banking in Knowledge City. Table S.3 summarizes the actions and policies to support the achievement of this goal.

Table S.3
Summary of Actions to Ensure the Availability of Innovation-Oriented Financing

Policy Area	Action	Priority	Ease	Sequencing
Innovation-oriented financing	Assist in the creation of angel investor networks. Entrepreneurs in GDD face challenges accessing early-stage financing. This will likely apply to entrepreneurs in Knowledge City. One way to fill this gap is to create formal networks of angel investors and give them the opportunity to regularly review promising investment opportunities.	High	Medium	Early

Table S.3—Continued

Policy Area	Action	Priority	Ease	Sequencing
Banking	Ensure the presence of competitive commercial banking in Knowledge City. Even if innovation-oriented financing is available, innovative firms will still need commercial banking services. GDD will need to ensure a competitive commercial banking presence in Knowledge City.	High	Easy	Middle
Innovation-oriented financing	Facilitate private venture investment. International experience shows mixed results with government-sponsored venture funds. Private venture capital has been better able to judge business prospects and risks. One way to involve private venture capital is by forming partnerships between foreign or domestic firms and government-sponsored firms.	Medium	Medium	Middle

Building on the Roadmap for Strategic Actions

This outline and the designated actions presented in Tables S.1–S.3 serve as guidance for a fully articulated strategic plan. That strategic plan should guide implementation by

- providing instructions for how the coordinated set of actions will be taken
- identifying who will take the actions
- setting out a timetable for the actions
- estimating the cost of the actions
- creating a full set of measurement indicators to evaluate outputs during the course of the planning and outcomes after the plan has been implemented.

This outline focuses on high-technology companies. However, it is important to recognize that such companies also need a variety of non-technology businesses to provide standard goods and services needed by all businesses; thus, a full strategic plan may need to differentiate between tasks to attract and support high-technology businesses and tasks for other types of businesses. Likewise, while we discuss various types of infrastructure that relate directly to increasing Knowledge City's potential to serve as a site of innovation, a full strategic plan should consider other infrastructure, including water, electricity, transportation, telecommunications, and sewerage. A strategic plan for Knowledge City will also need to be coordinated with urban development plans for surrounding areas.

The outline and the proposed strategy should serve as the basis for effective coordination and cooperation between GDD and Singbridge in implementing their marketing plans. Both have strong incentives to achieve success in Knowledge City. A strategic plan will be an outwardly visible signal of commitment by GDD that will help attract scientific, engineering, and entrepreneurial talent to Knowledge City.

Even with the best strategic plan, creating Knowledge City will be a long-term process that will require both consistent effort and flexibility in implementation. Continuity and consistency of policies will be particularly important. At the same time, the detailed strategy that will be the outgrowth of this outline should include mechanisms for reviewing policies and changing actions as the Chinese and global economies change and as Knowledge City develops.

Acknowledgments

Debra Knopman provided overall intellectual guidance, and Mu Dan Ping, a consultant for RAND, played an essential role in facilitating interactions between the RAND team and our GDD counterparts. Chaoling Feng provided essential research assistance and translation; we also benefited from the input of RAND colleagues Samuel Berkowitz, Kate Giglio, Scott Harold, Aaron Kofner, Alexandria Smith, and Bas Weerman. We greatly appreciate their efforts. The report was greatly improved by helpful and detailed reviews from C. Richard Neu at RAND and Sylvia Schwaag Serger, director of international strategy and networks at VINNOVA, as well as Wang Yuan and Wang Fenyu, executive vice president and vice president, respectively, of the Chinese Academy of Science and Technology for Economic Development. The report was much improved by the assistance of production editor Stacie McKee and editor Lauren Skrabala, who coordinated the publication process. We are also grateful to the many entrepreneurs, business people, investors, and government officials in GDD and Guangzhou, Silicon Valley, Maryland, Israel, and elsewhere who candidly shared information with us. Any errors are the responsibility of the authors.

Abbreviations

GDD	Guangzhou Development District
ICT	information and communication technology
IP	intellectual property
IPR	intellectual property rights
OECD	Organisation for Economic Co-operation and Development
R&D	research and development
TEDCO	Maryland Technology Development Corporation

CHAPTER ONE

Introduction

China's Guangzhou Development District (GDD) is focused on creating an environment conducive to innovation in a new development called Sino-Singapore Guangzhou Knowledge City (see Figure 1.1). Jointly developed by GDD and Singbridge of Singapore, Knowledge City is to be a new environmentally and technologically advanced city that hosts innovative industries and their associated knowledge workers. With a planned area of 123 square kilometers, much of which will be green space, Knowledge City will be larger than Vancouver, Canada (115 square kilometers), or Paris (105 square kilometers). The project is designed to spur the transformation of Guangzhou from less-complex manufacturing to higher-wage research- and innovation-based knowledge industries.

When completed, Knowledge City will be one of several components of GDD, a specially designated economic district in the Luogang District of Guangzhou. GDD includes Guangzhou Science City (also part of the Guangzhou High-Technology Industrial Development Zone), the Guangzhou Economic and Technological Development District, the Guangzhou Free Trade Zone, the Guangzhou Export Processing Zone, and Guangzhou International Biological Island. Figure 1.2 shows Luogang District, Guangzhou Science City, and the location of Knowledge City.

This document outlines a strategy for Knowledge City. It stems from research and analysis conducted by the RAND Corporation in collaboration with GDD between May 2011 and May 2012. The research effort consisted of three tasks. In task 1, we assessed existing models of innovation in GDD. In task 2, we analyzed the applicability of a small,

Figure 1.1
Proposed Knowledge City

SOURCE: Sino-Singapore Guangzhou Knowledge City.
RAND *MG1240-1.1*

select set of international best practices for GDD. Finally, in task 3, we developed this outline of a strategy for GDD, incorporating the research and analysis in the preceding tasks. The report's intended audiences include GDD and others interested in creating innovative areas.

This report is a companion volume to another RAND report, *Creating an Innovation System for Knowledge City* (Nataraj et al., 2012), a compilation of results from the interim analyses conducted for this project and supporting evidence for the conclusions presented here. The contents of that volume are listed in the appendix to this report. The intended audiences include GDD, economic development practitioners, and scholars of innovation and regional development.

Figure 1.2
GDD, Luogang District, and the Site of Knowledge City

SOURCE: Sino-Singapore Guangzhou Knowledge City.
NOTES: Knowledge City is the red area of Luogang District; the rest of Luogang District is shown in orange, and Guangzhou Science City is the area within the blue outline in the lower left part of Luogang District. Guangzhou City is further to the lower left, beyond Luogang District.
RAND *MG1240-1.2*

Goals of the Outline and the Strategy

For Knowledge City to be successful, GDD will need to create conditions that foster the creation of an innovative area to meet its three overarching goals:

1. Attract high-technology companies and enable their growth.

2. Attract and retain highly skilled, innovative people.
3. Ensure the availability of innovation-oriented financing.

The purpose of the strategy is to set GDD on a path toward creating conditions in Knowledge City that are conducive to innovation and the commercialization of new technologies. To this end, this outline of the strategy includes the following:

- a statement of GDD's challenges in meeting its goals
- proposed policies to address these challenges
- a coordinated set of actions to implement the policies and meet GDD's goals.

This outline also recommends priorities for the proposed actions, which are presented in order of their importance and ease of implementation. It provides a specific roadmap for GDD to follow in implementing the recommendations. In addition, it provides a set of indicators that will help measure changes in innovative capacity and activity.

This outline will serve as guidance for the fully articulated strategy to follow. That strategy should guide implementation by

- providing instructions for how the coordinated set of actions will be taken
- identifying who will take these actions
- setting out a timetable for the actions
- estimating the costs of the actions
- creating a full set of measurement indicators to evaluate outputs during the course of planning and outcomes after the plan has been implemented.

This report focuses on high-technology companies. However, it is important to recognize that such companies also need a variety of other businesses to provide goods, such as paper and office furniture, and to provide support services. Many of these companies may be small businesses. Some of our recommendations regarding high-technology companies will be applicable to these other businesses, but a full strategic plan may need to differentiate between tasks associated with high-technology businesses and tasks associated with other types of businesses.

Any successful city will need adequate infrastructure, including water, electricity, transportation, telecommunications, and sewerage. We discuss various types of infrastructure when they relate directly to increasing the chance of making Knowledge City a site of innovation. We have omitted other infrastructure issues, however; these considerations have been part of the overall master planning for the site, which is beyond the scope of our work.

A strategic plan for Knowledge City will need to take into account the planned development of surrounding areas. Therefore, it will be important to coordinate the development of a strategic plan for Knowledge City with other existing urban development plans. This coordination may entail not only tailoring the Knowledge City plan to existing plans but also updating the existing plans to account for development in Knowledge City.

Innovation Systems

Innovation does not happen in a vacuum. Researchers who focus on innovation have identified the presence of an *innovation system* as being important for innovation and its role in economic development (Organisation for Economic Co-operation and Development [OECD], 1997). Innovation systems consist of *actors* and the *connections* among them. Innovation policy can be defined as "a set of policy actions to raise the quantity and efficiency of innovative activities" (European Commission, 2000). Innovation systems and policies, while often focused on science and technology, also encompass social, political, and economic activities and institutions (Lundvall et al., 2002; Liu et al., 2011).

Over the past 30 years, China's innovation system has become more decentralized, and the country has made progress in developing many factors that support innovation (Liu et al., 2011; White, Gao, and Zhang, 2005; Yang, 2003). During the same period, China's innovation policies have become broader, moving from an exclusive focus on science and technology toward a framework that coordinates science and technology policies with industrial and fiscal policies (Liu et al., 2011). The country's current medium- to long-term plan for science and technology has three main goals: to increase research and development (R&D) expenditures to 2.5 percent of the gross domestic product by 2020, to shift toward "indigenous innovation," and to make the business sector the key force behind innovation (Schwaag Serger and Breidne, 2007).

At the provincial level, Guangdong is one of the top three regions in China in terms of total patent applications (Kroll, 2010). However, there are several challenges to Guangdong's innovation system, including scarce venture capital, weak implementation of intellectual property rights protections, and too few top universities and research centers; another potential concern is the concentration of R&D in the field of electronic and telecommunications equipment. To enhance its innovation system, Guangdong Province has developed a number of policies that focus on creating innovation networks, improving training and education, establishing research institutes, and implementing an intellectual property rights strategy, among other initiatives (Kroll and Tagscherer, 2009).

RAND has developed a simple structure for an innovation system in Knowledge City (see Figure 1.3). GDD's ultimate goal is for Knowledge City to emerge as a leading area of innovation. To meet this goal, GDD will need to attract innovative firms and enable their growth, attract and retain talented entrepreneurs and workers, and ensure the availability of innovation-oriented financing.

The necessary actions must be taken within an underlying legal and regulatory environment and the business support environment. One important goal for GDD is to make these environments as supportive as possible. When combined, these five elements will help foster the attraction, growth, and sustainment of innovative, high-technology firms in Knowledge City.

From the standpoint of RAND and GDD, the key reason to define and understand an innovation system is to find leverage points where policy measures may spur innovation. These leverage points may include government interventions regarding regulation, taxation, or

Figure 1.3
The Innovation System Framework

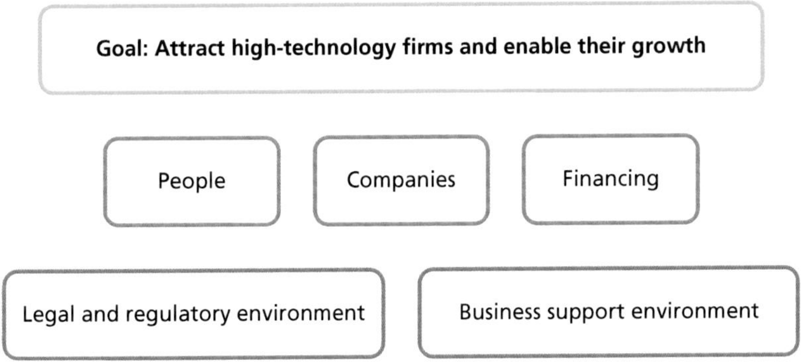

RAND *MG1240-1.3*

financing, for example, or they could involve interventions regarding how the different elements of the innovation system interact (OECD, 1997). The need to intervene could stem from a failure of the market, but intervention could also involve changing or removing a government policy—in effect, correcting a government policy failure. These interventions collectively amount to innovation policy.

Knowledge City will be successful if it becomes home to a cluster or clusters of innovative industries. Clusters may be described in a number of ways; one of the more commonly used definitions is from Michael Porter (1998), who defines clusters as "geographic concentrations of interconnected companies and institutions in a particular field." He notes that clusters may include not simply one industry but "an array of linked industries," including suppliers and customers.

The emergence of clusters is a sure indicator that an environment has been created that favors innovation. Clusters emerge from the assets of the region, policies designed to improve those assets, and policies to improve the innovation environment. The actions summarized in this outline of a strategy constitute elements of all five building blocks in the innovation system framework that are needed to attract high-technology firms and clusters and enable their growth.

Methods

RAND and GDD used a variety of methods to develop this outline of a strategy.

- *Interviews.* We conducted two sets of interviews. First, we met with business executives in high technology and other sectors in GDD and Guangzhou to understand the landscape for innovation and the specific business challenges they face. Most of our interviewees were selected with the assistance of GDD. Next, we met with specific individual international investors who are active in China to obtain their informed outsider's perspective on the opportunities and challenges GDD will face in developing Knowledge City. We selected the international investors based on personal connections.
- *Literature reviews.* We reviewed the literature on business conditions in China, innovation in China, and the growth of innovation areas around the world.
- *Data analysis.* We received from GDD data on economic and demographic conditions in GDD and Guangzhou, and we compiled data on likely competitor regions in China. Our purpose was to determine baseline conditions for innovation in Guangzhou and elsewhere.
- *Case studies.* In consultation with GDD, we selected three innovative areas as the targets of case studies, the purpose of which was to identify international best practices and apply them to Knowledge City. Our case studies included the Silicon Valley area in California, the life sciences corridor in Maryland, and the technology industries in Israel. For each case study, we reviewed relevant literature, compiled and analyzed data about the area in question, and talked either by telephone or in person with key participants in the growth of these areas.
- *The GDD-RAND Knowledge City Innovation System Research Project Survey.* RAND and GDD jointly conducted a survey of high-technology firms already in GDD. RAND and GDD designed the survey instrument, and GDD staff fielded the survey and

entered the data. RAND staff then analyzed the data. The survey was designed to elicit responses about all aspects of the innovation environment in GDD.

Goals, Attributes for Success, and Priority Actions

To achieve its goal of creating Knowledge City as a new environmentally and technologically advanced city that hosts innovative industries and their associated knowledge workers, GDD must convince innovative companies and talented individuals to stake their futures on an unproven new development. To convince them, GDD must instill confidence in potential partners that it has the attributes that will result in the successful creation of an innovation-based cluster.

These attributes fall into three broad categories. The first encompasses the natural advantages of an area. Such advantages are innate and cannot be replicated by other areas. For example, in our case studies, Silicon Valley has the natural advantage of the mild weather of Northern California, and Maryland is home to the National Institutes of Health. The second category entails factors that can be easily replicated. According to our findings from the case studies, tax concessions and non-tax incentives fall into this category: Almost any site can replicate these incentives. The third category consists of factors that may eventually be replicated, but doing so would take considerable time and effort. This category includes high-quality infrastructure, a good business climate, a reputation for intellectual property rights (IPR) enforcement, strong angel investor networks, and excellent quality of life and local schools.

Focusing on these factors creates a competitive advantage. First, it takes time to create these types of institutions, making it harder for other regions to compete. Second, clusters are often formed around the area where the industry began. The original site of an industry creates a first-mover advantage. Although this advantage may be due to historical accident, having attractive assets increases the probability that a new product or service will be developed in Knowledge City, creating the basis for the formation of a new cluster.

To instill confidence, GDD will need to emphasize the attractiveness of its existing assets to advanced companies and talented individuals. To increase its chances of success, GDD will also want to take steps to create assets that other regions will find hard to replicate. Creating a reputation for having these assets could assist in attracting innovation-oriented firms to Knowledge City; the presence of these firms would reinforce Knowledge City's reputation for innovation, thus attracting more innovation-oriented firms to the area and creating a virtuous circle that would make it difficult for other regions to catch up.

The proposed strategy can also serve as the basis for effective coordination and cooperation between GDD and Singbridge in implementing their marketing plans. Both have strong incentives to achieve success in Knowledge City. A strategic plan will be an outwardly visible signal of commitment by GDD that will help attract scientific, engineering, and entrepreneurial talent to Knowledge City.

We present this outline of a strategy in five sections:

1. actions to attract high-technology companies and enable their growth
2. actions to attract and retain highly skilled, innovative people
3. actions to ensure the availability of innovation-oriented financing
4. ranking of these actions by priority, ease of implementation, and order of implementation
5. identification of indicators through which GDD can assess progress in the creation, growth, and sustainment of high-technology firms.

Even with the best strategic plan, creating Knowledge City will be a long-term process that will require both consistent effort and flexibility in implementation. Continuity and consistency of policies will be particularly important. At the same time, the detailed strategy that will be the outgrowth of this outline should include mechanisms for reviewing policies and making changes as Knowledge City and as the Chinese and global economies change. Part of this review should include an assessment of whether investors and businesses are succeeding and the extent to which Knowledge City policies and programs have enabled this success.

CHAPTER TWO

Attracting High-Technology Companies and Enabling Their Growth

This chapter presents actions for GDD to pursue to attract high-technology companies and enable their growth. These actions can be grouped into two domains:

- attracting an anchor institution and marketing Knowledge City
- improving the overall environment for innovation rather than only targeting specific sectors.

This chapter briefly reviews and assesses GDD's assets and the assets that Knowledge City is likely to have that will help it attract and retain high-technology companies. It then provides details on the two domains and outlines the actions that GDD should take to capitalize on its assets and improve its ability to attract and retain high-technology companies.

GDD's Assets

GDD and the Guangzhou area have a number of assets that will be attractive to high-technology companies. The Guangzhou area has served as a center of global commerce for generations, an ethos that remains. Advanced, high-throughput seaports and Guangzhou's major airport are important assets; they are tangible signs of Guangzhou's openness to the world. These transportation hubs provide quick links to the dynamic Southeast Asia region and to the rest of the world. Further enhancing its connections, the area also has excellent intraregional transport infrastructure. This transportation infrastructure will be available to Knowledge City (see Figure 2.1).

The Guangzhou area has assets directly related to knowledge generation—specifically, two universities among the top 50 in China: Sun Yat-Sen (Zhongshan) University (ranked 16) and South China University of Technology (ranked 29) (Shanghai Ranking Consultancy, 2011). An alternative university ranking system implemented by a team at the Chinese Academy of Management Sciences ranks these universities even higher, with Sun Yat-Sen University at number 7 and South China University of Technology at number 24 ("Top 100 Chinese Universities in 2011," 2011). The team at the Chinese Academy of

Figure 2.1
Knowledge City's Location at the Heart of the Pearl River Delta

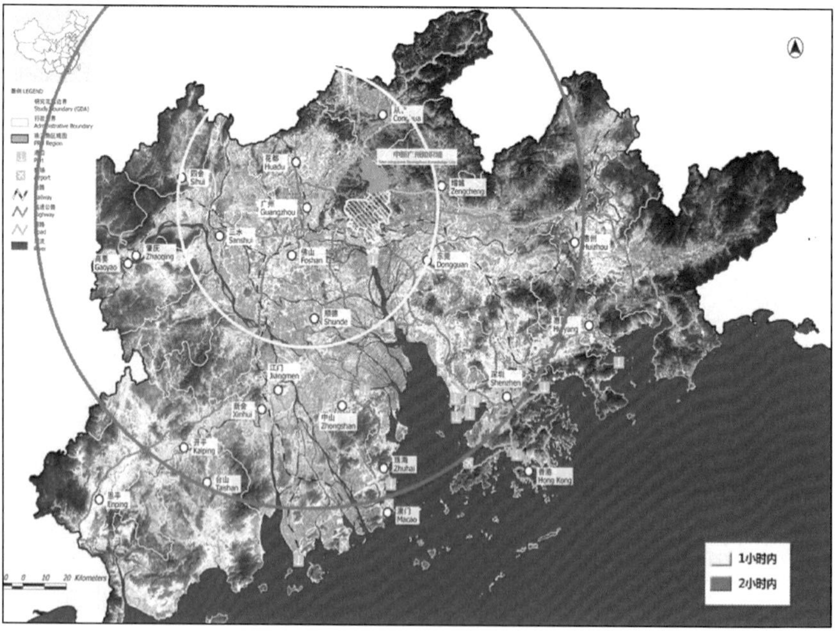

SOURCE: Sino-Singapore Guangzhou Knowledge City.
NOTE: Knowledge City is being developed in the orange area on the map. It is part of Luogang District, the other part of which is the area outlined in yellow. The yellow and blue circles indicate the areas within one and two hours' driving distance of Knowledge City, respectively.
RAND *MG1240-2.1*

Management Sciences also ranks Sun Yat-Sen University as eighth in science, and South China University of Technology as ninth in engineering. These rankings suggest that the two universities may provide a solid basis for carrying out basic research that can feed into commercial product and process innovations. One potential mechanism is institutional collaboration between faculty members and innovative companies, as demonstrated by the GDD Zhongshan Biotech Industrial Research Institute.

GDD's existing base of companies provides an excellent starting point for identifying and attracting high-technology firms to Knowledge City. The GDD-RAND Knowledge City Project Survey identified four groups of companies that have a major presence in GDD: companies engaged in manufacturing electronics and information technology (41 percent of 305 responses), biological and pharmaceutical firms (22 percent), manufacturers of new materials (7 percent), and companies involved in the optical, mechanical, and electronic integration sectors (7 percent).[1] These responses suggest that GDD has a comparative advantage in these sectors. Knowledge City may therefore wish to build on this comparative advantage.

Attracting an Anchor Institution and Marketing Knowledge City

Our case studies of Silicon Valley, the life sciences cluster in Maryland, and the high-technology cluster in Israel, along with the literature on technology clusters, emphasize the important role played by a major company or institution in creating a nucleus around which a cluster can form. We call such a company or institution an *anchor institution.* We recommend that GDD focus its marketing efforts on attracting an anchor institution or a small set of anchor institutions to Knowledge City.

---

[1] The GDD-RAND Knowledge City Project Survey methods and results are described in the companion RAND report, *Creating an Innovation System for Knowledge City* (Nataraj et al., 2012).

Anchor institutions could include the following:

- high-profile companies with substantial R&D facilities
- research institutes
- research universities.

One type of anchor institution is a high-profile business that would site its R&D activities within Knowledge City. Even better would be a business that places its Chinese headquarters, production facilities, or marketing and logistics departments, as well as its R&D activities, in Knowledge City.

Two other options are major nonprofit research facilities or research-oriented universities that employ teams of postdoctoral and Ph.D.-level researchers working on advanced problems. Research institutes or universities can contribute to the formation of a cluster by transferring their discoveries to local firms, by adapting knowledge created elsewhere to local conditions, and by attracting new human, knowledge, and financial resources. Unfortunately, these institutions are not self-supporting: If Knowledge City is to attract such an institution, the institution will need national or local funding for its activities. If Knowledge City attracts a nonprofit research institute or university, GDD will want to ensure that the institution is closely linked to local businesses and that it sets up effective procedures and incentives for encouraging staff to commercialize their research.

Each of the clusters we studied was formed around the presence of one or more anchor institutions. In the case of Silicon Valley, Stanford University, Fairchild Semiconductor, and the federal government through its role as a major purchaser of technology all served as anchor institutions at one time or another. In Maryland, the National Institutes of Health and other federal laboratories and agencies have filled the role of anchor institutions, providing a large pool of highly skilled workers as well as demand for services from growing life sciences firms. In the case of Israel, our interviews suggest that the culture of universal conscription, and the emphasis placed by the military on the development of new technologies, made the Israel Defense Forces the original anchor institution. Subsequently, multinational corporations, such as

Intel and Microsoft, also played important roles as anchors by training and employing thousands of information and communication technology (ICT) workers and creating ties between Israel and global industry.

An anchor institution provides a number of benefits:

- It attracts suppliers and customers with which it wishes to do business.
- It serves as a source of talent.
- It serves as a source of research and spinoff companies.
- If it is a major company, it provides a bridge to global markets and training for entrepreneurs in accessing these markets.
- Its presence sends a signal to other firms and individuals that Knowledge City is a desirable location.

Conditions in GDD

GDD's assets have helped it develop a strong manufacturing base with great experience in exporting. This base is currently a source of incremental innovation—continuous improvement in products or processes. Representatives from one industry group indicated in our interviews that they recently collaborated with a university to increase the productivity of part of their manufacturing process. Another interviewee noted that companies in Guangzhou and south China have become adept at adapting their products for different market niches in the country.

In some respects, Guangzhou's strong manufacturing base may be a disadvantage. Because of its success in manufacturing, investors have voiced some skepticism that the area can be a site for advanced product and technology innovation; GDD is associated primarily with assembly operations, not the development of new, innovative products. Despite Guangzhou's experience with market-driven businesses and international openness, several interviewees noted that the entrepreneurial culture in developing new product innovations is lacking. An R&D-oriented anchor institution can help dispel these perceptions.

More broadly, an anchor institution can highlight the attractiveness of Guangzhou compared to other regions of China. Some of Guangzhou's assets are similar to those in such competitor cities as

Beijing, Shanghai, Hangzhou, Tianjin, Suzhou, and Shenzhen. In our survey, technology executives in GDD considered the Yangtze River Delta, including Shanghai, the most suitable location for their business after GDD (see Figure 2.2). Next-highest ranked were other cities in Guangdong Province. These responses reflect the regions against which GDD will need to compete to attract innovative firms.

An anchor institution that is a business can help attract suppliers and, if it were part of a multinational business group, could expand the number of marketing channels from Guangzhou to the world. Company officials noted in the GDD-RAND Knowledge City Project Survey that GDD has an extensive supplier base and inexpensive support services, such as those provided by lawyers and accountants. A business anchor institution would foster the development of this supplier base and these support services. Several interviewees mentioned that GDD's policies have helped them build their supplier, buyer, and support networks. At least one experienced China investor praised Guangzhou as one of the few places in the country with a strong cohort of career managers—people who do not necessarily aspire to own their own businesses but are experts in managing businesses. However, some

Figure 2.2
Firms' Perceptions of Suitable Locations Other Than GDD

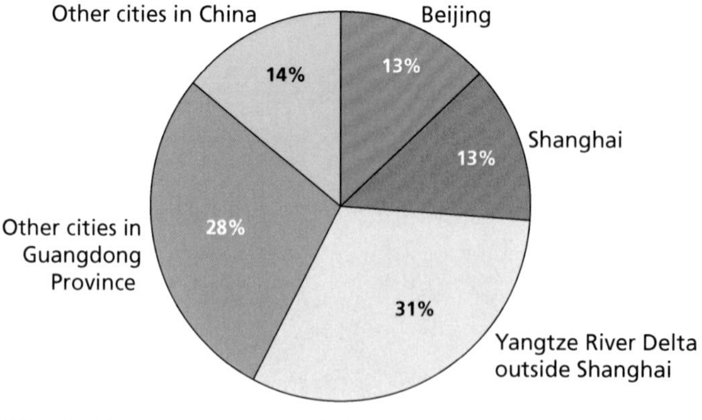

SOURCE: The GDD-RAND Knowledge City Project Survey.
NOTE: Results based on responses from 267 firms (out of 305 surveys received). Percentages do not sum to 100 due to rounding.
RAND *MG1240-2.2*

entrepreneurs mentioned difficulties in finding local suppliers and indicated that marketing channels were sometimes complicated.

Although GDD and Singbridge should first focus on attracting an anchor institution, they will need to continue to make other prospective companies aware of the opportunities in Knowledge City. Creating a professional marketing effort is central to this effort.

At this time, both GDD and Singbridge are targeting companies in a large number of industrial sectors (see Table 2.1). Experienced international investors noted to us that it is difficult for economic development authorities to tailor their policies to a large number of industrial sectors. If GDD chooses to target sectors, it might increase its chances of success by narrowing the range. Targeting fewer sectors will help focus GDD's marketing message about Knowledge City. However, our research suggests that GDD is likely to have even greater success if it focuses on improving the general business environment rather than only targeting specific industrial sectors. The policies that GDD implements will also need to be explained to prospective investors.

Table 2.1
Comparison of Target Industries Identified by GDD and Singbridge

GDD Investment Promotion Office September 21, 2011	Singbridge Presentation April 11, 2011
Next-generation information technology	Information technology convergence
Biotechnology and health	Biotechnology and pharmaceuticals
New energy, energy saving, and environmental protection	Energy and clean tech
New materials	Advanced manufacturing
Science and technology services	R&D services
Culture and creative industries	Creative industries
Headquarters siting	Education and training
	Health and wellness

Actions for GDD

The role that government policy can play in attracting anchor institutions is a subject of extensive debate. One concern with attempts to attract such institutions is that local governments often compete with one another by increasing the size of incentives offered to the institution until the costs outweigh any potential benefits (see, for example, Porter, 1996). However, Greenstone, Hornbeck, and Moretti (2010) suggest that, despite the risk of sparking a bidding war, there may be benefits to attracting large anchor institutions to a local area. The authors compared various outcomes in counties in the United States that succeeded in winning what they termed *million-dollar plants* to counties that were short-listed by the plant owners but were not chosen.[2] They found that the productivity of incumbent firms in winning counties, as well as county-level wages, increased in the years following the openings of million-dollar plants relative to the productivity and wages in losing counties. However, the authors cautioned that there was a substantial amount of variation in the results, with some winning counties experiencing large productivity gains but some experiencing productivity losses.

GDD should take the following actions to attract an anchor institution:

- First, focus on businesses already in Guangzhou. Several multinational companies are already located in the region. They are familiar with business practices in the Guangzhou area and already have a network of local suppliers and customers. They are part of broader global supplier and buyer networks and can supply leads to promising potential anchor institutions. These multinationals can serve as anchor institutions in several ways if they relocate to Knowledge City:
  - increasing the amount and complexity of their R&D effort
  - establishing training centers in their Knowledge City facilities

---

[2] The term *million-dollar plant* refers to plants listed in a regular feature in the corporate real estate journal *Site Selection* (Greenstone, Hornbeck, and Moretti, 2010).

  - adding or increasing existing production functions if they are linked to R&D
  - establishing Chinese or Asian headquarters, if they do not have one already
  - encouraging suppliers to locate in Knowledge City
  - encouraging key customers to locate branches in Knowledge City to work with them on developing new products.
- Include manufacturing firms in the search. Most of GDD's high-technology companies are engaged in manufacturing or are linked to manufacturing. Interviews suggested that manufacturing firms in Guangzhou have been quite good at incremental process and product innovation.
- Involve senior Guangzhou City or even Guangdong Province officials early in the search. To attract a highly regarded anchor institution, GDD will need to show that the local and provincial governments strongly back Knowledge City. Although staff can lay the groundwork and prepare background and briefing materials, high-level contacts will be necessary early on to signal local and provincial support.
- Be prepared to create an ample incentive package. Such a package could include reduced-price or free land, facilities, and infrastructure for the institution. However, offering incentives can lead to bidding wars with competitors that may not pay off for Knowledge City. In addition, designing such a package should include an analysis of potential benefits to reduce the possibility of making too generous an offer. That said, landing a major facility is likely to have substantial long-term benefits. Providing an incentive package for a key institution will likely generate large returns.
- The involvement of senior officials and the preparation of an incentive package will also be necessary steps for attracting a research institution or a university. In those cases, institutions may require living quarters for students, postdoctoral scholars, and faculty, or they may require specialized research facilities.

Attracting an anchor institution should be part of a broader effort to market Knowledge City. In light of the intense international competition to attract high-technology businesses to new developments, Knowledge City needs to be marketed in the most effective way possible. We recommend two specific marketing actions for GDD:

- Hire a specialist international marketing or public relations firm that has experience in (1) developing a marketing message for business parks aimed at an international audience, (2) encouraging companies to relocate to these parks, and (3) recruiting entrepreneurial talent. The firm should also help coordinate the message between Singbridge and GDD. Global competition among new cities focused on innovation is ferocious. An expert marketing firm can provide an edge for GDD.
- Sponsor an annual international trade fair or conference in an industry that is germane to Knowledge City. This could be the industry in which the anchor institution is engaged or an industry in which GDD currently has large numbers of companies, such as electronics and information technology, biotechnology, advanced materials, or optical, mechanical, and electronic integration. The purposes of such a trade fair would be to highlight Knowledge City as a location for the industry and to create connections between Knowledge City businesses and suppliers and customers from around the world. It is important that the trade fair be international, so that there is an international exchange of information, and that it is held in Knowledge City to maximize local attendance. Examples include the International Motor Show in Hannover, Germany, and the International Consumer Electronics Show in Las Vegas.

Improving the Overall Innovation Environment

There is an ongoing debate about whether government policy should selectively promote specific sectors. A number of commentators have argued that one of the most important initiatives governments can take to promote cluster formation and growth is to improve the overall business conditions for all firms—for example, by protecting IPR, developing infrastructure, investing in a skilled labor force, or lowering the cost of starting a firm (see, e.g., Hospers, Sautet, and Desrochers, 2008; Porter 1998). Evidence has suggested that "governments are not particularly good at picking winners" and that industrial policy—selectively targeting certain industries—may create industrial lobbies (Aghion et al., 2011).

However, some authors have argued that industrial policy can be designed to effectively promote high-growth sectors. Recently, Lin (2011) and Lin and Monga (2011) have argued that countries can succeed in targeting industries through certain policies if they follow specific criteria (outlined in Lin and Monga, 2011) and if those industries match the countries' comparative advantages.

We recommend GDD shift its emphasis more toward improving the general innovation environment rather than just targeting specific sectors. Although we do not rule out targeting sectors, truly innovative companies could come from any sector and merit support, given the goals of Knowledge City. Physical infrastructure constitutes one type of asset that would improve the overall environment. Another asset is the policy infrastructure, or the laws, regulations, and government actions that support innovation.

Conditions in GDD

We focus on three broad policy areas in which to improve the overall innovation environment: taxes, non-tax incentives, and IPR. We also discuss specific issues related to the size of the market.

Taxes

Overall, the tax situation in GDD does not appear to impose a major constraint on innovation. Most tax policies are set at the national or provincial level, thus allowing little scope for GDD to affect taxes on firms operating within its jurisdiction. In the area of tax policy, our interviews suggested that the most important benefit provided by GDD appears to be assistance with certifying that a resident company is eligible for national tax preferences and working with the tax administration to ensure that the company receives those benefits.

Non-Tax Incentives

GDD provides a number of non-tax incentives. The ability to provide cost-competitive land, office, and production facilities appears to be quite favorable for innovation. Grants for R&D, money to support inter-institutional collaborations, and softer aid for innovation, such as help with marketing and networking, were also regarded favorably, but their effectiveness was uncertain.

Intellectual Property Rights

Weak IPR can directly harm innovators and innovation by lowering the incentives to innovate. Insufficient IPR protection is associated with lower levels of investment in high-technology industries and more investment in distribution and simple assembly, along with lower levels of participation in advanced global research networks.

The patenting process in China appears to be relatively simple and efficient: Eighty-three percent of GDD-RAND Knowledge City Project Survey respondents indicated that they patent their innovations to protect them (see Figure 2.3). However, our interviews suggested that enforcing patents and otherwise protecting IPR remains a challenge, particularly for firms in industries such as software. This is echoed by several studies that have documented that many firms find it difficult to enforce their IPR in China, even though China has improved its legal IPR framework (see, e.g., Keupp, Beckenbauer, and Grossman, 2009; Quan and Chesbrough, 2010).

GDD has already taken steps to help firms with patenting and enforcing patents, including setting up a high-level IPR service center in Science City, one of several units of GDD. However, currently, only 4 percent of firms in the survey reported that they consult the government if their patents are infringed (see Figure 2.3). However, 40 percent indicated that GDD administration does provide help in enforcing IPR or patent rights, and 80 percent indicated that it was easy to obtain services related to IPR.

One other noteworthy fact about patenting in China is that very few Chinese-origin patents are granted internationally (World Intellec-

Figure 2.3
How GDD High-Technology Firms Protect Intellectual Property

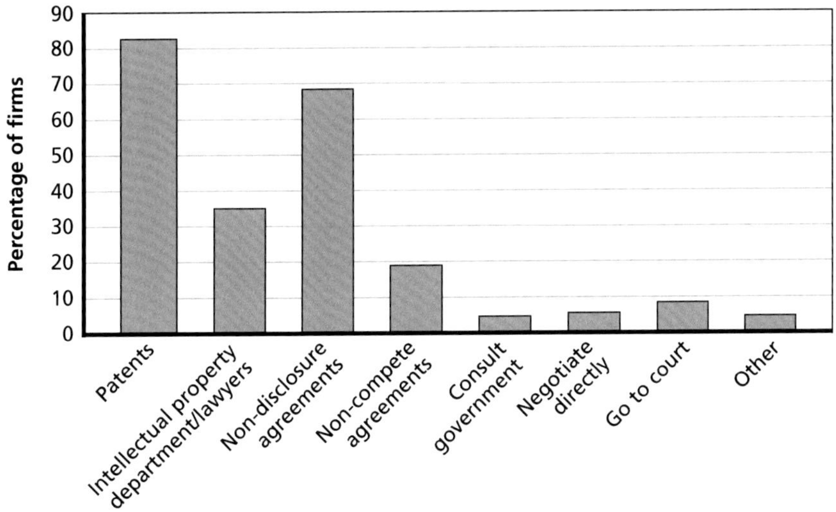

SOURCE: GDD-RAND Knowledge City Project Survey.
NOTE: Firms could give more than one answer to this question, so the total sums to more than 100 percent. Results are based on responses from 301 firms (out of 305 surveys received).
RAND *MG1240-2.3*

tual Property Organization, 2011). Approximately 5 percent of all Chinese-origin patents are registered outside of China. In contrast, more than one-quarter of U.S.- and Japanese-origin patents are granted outside their home countries, although the shares differ from year to year. Evidence suggests that the innovations for which patents are being granted in China might not be as original as they would need to be to qualify for patents from advanced-country patent offices (Liang, 2012; Cyranoski, 2010; "Patents, Yes; Ideas, Maybe," 2010). Reasons include lower standards in China, the fact that one can patent minor improvements that might not qualify for patents elsewhere, and the low costs and strong incentives for filing patents (Liang, 2012). China's *National Patent Development Strategy (2011–2020)* calls for the filing of more foreign patents (China State Intellectual Property Office, 2010).

Size of the Market

The Pearl River Delta and Guangdong Province are enormous markets, and our interviews indicated that they may seem an easy outlet for innovative products in the short term. However, interviewees indicated that local markets may not be optimal for innovation. Despite the size of these markets, we were told that consumers lack the levels of sophistication found in markets in wealthier countries, especially those in Japan and the Republic of Korea. Demand in these markets for innovative high-technology products and services drives innovation. Local markets may not provide firms with the experience to penetrate challenging markets, and high-technology companies have to establish themselves in the most competitive markets for long-term success. GDD will need to work to ensure that high-technology companies in Knowledge City have access to the most demanding markets for their products and services.

Actions for GDD

To improve several aspects of the overall innovation environment, GDD should take the following actions:

- Create and maintain an office to assist companies with all of their legal, administrative, and financing issues. Such an office should have records of all relevant laws and regulations, all incentive programs, and all application forms for any benefits for which a company can apply. It should be charged with assisting companies in understanding and taking advantage of these programs.
- Assign each new company a case officer from this office who is responsible for settling the company in Knowledge City. Case officers would help companies understand the various applicable taxes, licenses, and regulations. They would help the companies obtain licenses, permits, and tax incentives for which they qualify at the GDD, municipal, provincial, and national levels. Case officers would continue to assist companies as they grow and would help resolve any disputes or problems with government offices.

Taxes

In some countries or regions, higher personal, corporate, and capital gains taxes appear to have led to lower levels of entrepreneurship. However, evidence from our case studies indicates that taxes are less important for attracting innovative companies than labor quality, proximity to suppliers and customers, and quality of life. Our interviews and the GDD-RAND Knowledge City Project Survey provided little indication that the current tax system in GDD constrains innovation.

GDD has tools to address the specific tax-related concerns of potential tenants. In some instances, GDD may need to address these concerns directly, such as when attracting an anchor institution. However, GDD is likely to face competition from other regions in China concerning tax incentives. Therefore, we recommend the following actions:

- Charge the company-assistance office with helping companies that are new to GDD understand the various taxes that are applicable to them. The case officer assigned to the company should help it obtain any tax incentives for which it qualifies at the municipal, provincial, and national levels. The office would also monitor tax incentives provided by competitor regions.
- When necessary, offset taxes with a package of non-tax incentives. GDD has limited room for maneuver on taxes, since most policies are set at the national level.

Non-Tax Incentives

Non-tax incentives include assistance with obtaining land, office or industrial space and equipment, space in incubators, and preferential leasing, among other incentives. Interviews with innovative firms in GDD indicate that such non-tax incentives are helpful. However, as with taxes, GDD will always face competition from other economic zones in providing non-tax incentives. Therefore, the design of incentive packages should attempt to account for likely benefits as a way of determining the appropriate size of any package. We recommend that GDD pursue the following initiatives:

- Request that the company-assistance office keep track of municipal, provincial, and national non-tax incentives and assist firms in obtaining those incentives. Such an office should include a database of all possible incentives, as well as instructions for applying and the forms to use to obtain these incentives, including municipal, provincial, and national R&D grants. It should help firms obtain these incentives by helping them fill out these forms and by working with other government offices to ensure that companies receive the incentives for which they qualify.
- Ensure that non-tax incentive packages, such as grants, loan subsidies, or direct purchases by GDD on behalf of firms, can also respond to the unique needs of innovative businesses. Although assistance with land and buildings can always help, non-tax incentives should also include help with making expensive technical equipment or laboratory space available to firms.

Intellectual Property Rights

If GDD wants Knowledge City to succeed as a location for advanced technology and innovation, strong IPR protection must play a role. GDD can take two actions to promote IPR:

- Make Knowledge City a zone of IPR enforcement from which IPR cases will be pursued aggressively throughout China. The GDD-RAND Knowledge City Project Survey indicated that many technology firms in GDD rely on patents to protect their intellectual property. GDD can differentiate itself from other development zones by proving to domestic and international investors that the local government will assist firms in Knowledge City in protecting their intellectual property throughout China. We recommend that GDD work with the Guangzhou City government to set up a dedicated office to pursue IPR cases throughout China.
- Expand assistance for international patenting. Helping Chinese inventors and innovators with international patenting processes could provide an incentive for the creation of higher-quality intellectual property than would be the case with just patenting in China.

Expanding Markets

Research has shown that market demand can provide a strong incentive for innovation. Technology firms in Knowledge City will benefit from having to meet the high standards of wealthy and developed market economies. Two nearby markets are wealthier than Guangdong Province: Hong Kong and Taiwan. Guangzhou is already strongly linked to Hong Kong; the national government and the Guangdong provincial government have plans to strengthen these links. Many investors from Taiwan have established businesses in Guangzhou. Further afield, but still easily accessible geographically, the Republic of Korea and Japan have demanding customers to whom Knowledge City firms should aspire to sell. We recommend that GDD take the following action in this area:

- Support initiatives by high-technology firms in Knowledge City to market their products in Hong Kong and Taiwan. GDD may wish to strengthen representative offices, encourage entrepreneurs to attend trade fairs in each location, and sponsor special events. Although the Korean and Japanese markets are further away and may have better links to Shanghai, Beijing, and Tianjin, these markets should also be considered for special targeting.

We noted earlier that our analysis of the innovation environment focused on taxes, non-tax incentives, and IPR—three broad policy areas that are part of the legal and regulatory environment and the business support environment, the two basic building blocks of the innovation system framework (see Figure 1.3 in Chapter One). Part of the legal and regulatory environment also encompasses the natural environment, with appropriate standards for such issues as environmental compliance. Given the environmental goals of Knowledge City, GDD will want to develop standards for air, soil, and water pollutants. These could take the form of maximum allowable emissions from a facility, or they could be determined on a case-by-case basis. Similarly, GDD may want to develop minimum standards for water use efficiency. These regulations and standards could serve as barriers to some types of industries, but they also could make Knowledge City more attractive to industries and workers who value or need a cleaner environment. Such regulations and standards should be incorporated into a full strategic plan.

Summary of Actions for Attracting High-Technology Companies and Enabling Their Growth

The actions suggested in this chapter will work together to create an environment for innovation in Knowledge City. Attracting an anchor institution will help induce other companies to set up in or move to Knowledge City. Strong IPR enforcement, when it is within the power of GDD, and assistance with enforcement otherwise will clearly differentiate Knowledge City and GDD from other regions in China, providing a strong incentive for innovative companies to locate in Knowledge City. Tax and non-tax incentives will lower the cost of establishing a business in Knowledge City. Helping high-technology companies located in Knowledge City to explore new markets with high standards for innovation and quality can help ensure their long-term success. Finally, marketing Knowledge City properly will help attract prospective investors and make prospective customers aware of the products and services being produced in Knowledge City.

CHAPTER THREE

Attracting and Retaining People

This chapter recommends actions for GDD to pursue to attract and retain highly skilled, innovative people. The actions can be grouped into two domains:

- actions that enhance the quality of life in Knowledge City
- actions that attract talent and build networks.

GDD's Assets

GDD and Guangzhou have been attractive locations for the types of people who will establish and work in the innovative companies of Knowledge City. GDD has a base of serial entrepreneurs, people with experience starting technology companies. In the GDD-RAND Knowledge City Project Survey, almost half of respondents said that they had previously started a company. The majority of companies responding to the GDD-RAND Knowledge City Project Survey were founded as independent companies in either GDD or Guangzhou, rather than as subsidiaries of other companies.

Guangzhou has two of the top 50 universities in China (see Figure 3.1). Universities are important to an innovative area, in part because of their role in training and educating the workforce. Guangzhou's universities and technical schools have trained workers in the specialties needed by high-technology companies. The vast majority of companies in the GDD-RAND Knowledge City Project Survey said that they had little difficulty finding the staff they needed locally,

Figure 3.1
Sun Yat-Sen University and South China University of Technology, Two of the Top 50 Universities in China

SOURCES: Sun Yat-Sen University and South China University of Technology. Used with permission.
RAND *MG1240-3.1*

although they still needed to train the staff once hired. In addition, GDD boasts about 2,000 returnees from overseas and about 700 companies founded by returnees. According to our research, immigrants and their networks with their countries of origin have been very important for the creation of successful clusters, including the Israeli technology cluster and the Hsinchu cluster in Taiwan.

Quality of Life

The availability of skilled labor is generally more important than tax and non-tax incentives for ensuring the success of a cluster. Skilled labor will come to or stay in GDD for one of two reasons. The first is employment opportunities. The second is quality of life.

In all three of our case studies, quality of life was an important factor. Kolko, Neumark, and Mejia (2011) have argued that California's mild climate and dry weather could explain much of Silicon Valley's growth. In Maryland, a number of observers have indicated that quality-of-life issues, such as excellent public schools, urban amenities, and a good transportation system, were key criteria that attracted them or their firms to the area. For example, one official recounted that a major company that recently relocated to the cluster chose its location because of good public transit and urban amenities. In Israel, the main hubs of the ICT cluster (Herzliya, Tel Aviv, Ra'anana, and Haifa) are considered the most desirable places to live in the country.

Quality-of-life issues are often rated highly in other clusters. Athreye (2004) documented the results of a firm survey in the ICT cluster in Cambridge, United Kingdom, in which the factor most often ranked as important for the firm's development was an "attractive local living environment for staff/directors." Dahl and Sorenson (2010) highlighted a related issue, showing that proximity to friends and family plays a large role in the location decisions of Danish scientists and engineers.

Conditions in GDD

GDD does face some challenges in the area of quality of life. First, although most companies can find the talent they need, some cannot. Most of the representatives we interviewed said it was particularly difficult to attract top talent from Shanghai and the Yangtze River Delta, which we found to be GDD's main competitor area. Residents of Beijing were also reluctant to move to Guangzhou. Retention of top talent can also be difficult. Quality of life is related to these challenges.

During our interviews in GDD, entrepreneurs and investors mentioned a shortage of low-cost housing; a lack of local amenities, such as schools, hospitals, and shopping centers; and a long commute from Guangzhou City, where many people prefer to live or where one spouse often works. Among survey respondents, the issue most frequently ranked among the top two challenges in recruiting and retaining staff was reported to be salary. However, quality-of-life issues were also critical. As shown in Figure 3.2, after salary, the most frequently cited difficulties in recruiting staff were commuting time (40 percent), reluctance to live in GDD (24 percent), housing prices (12 percent), company reputation (11 percent), family reasons (10 percent), and educational opportunities for children (8 percent).

Figure 3.2
Most Frequently Cited Reasons for Difficulty in Recruiting Staff

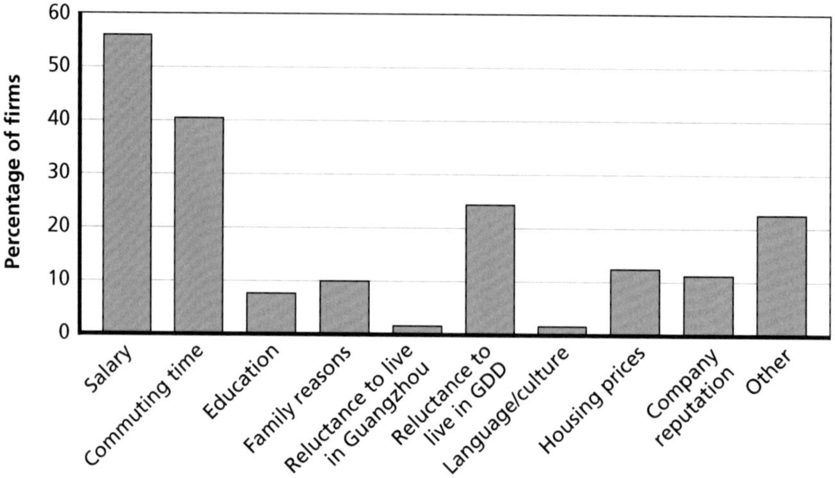

SOURCE: GDD-RAND Knowledge City Project Survey.
NOTE: Firms could give up to two answers to this question, so the total sums to more than 100 percent. Results are based on responses from 301 firms (out of 305 surveys received).
RAND *MG1240-3.2*

Similar difficulties were cited for retaining employees. These challenges, particularly the commuting issue, will be exacerbated in the case of Knowledge City, which is located in a part of Luogang District that is far from Guangzhou City (see Figure 3.3). During our interviews, several entrepreneurs mentioned GDD policies that have helped to mitigate the land and housing challenges, including GDD's programs to build dormitories in the area and to attract a real estate developer to build apartments. However, one entrepreneur noted that despite such efforts, the rent for such apartments remains unaffordable for many.

Our review of the literature and our case studies suggest that quality of life plays a role in determining where highly skilled people choose to live. Our survey of high-technology firms confirmed that quality-of-life issues, such as commuting and, more broadly, a "reluctance to live in GDD," are potential impediments to attracting top talent. Although housing prices can indeed be a problem, people will pay a premium to

Figure 3.3
Location of Knowledge City Relative to Guangzhou City

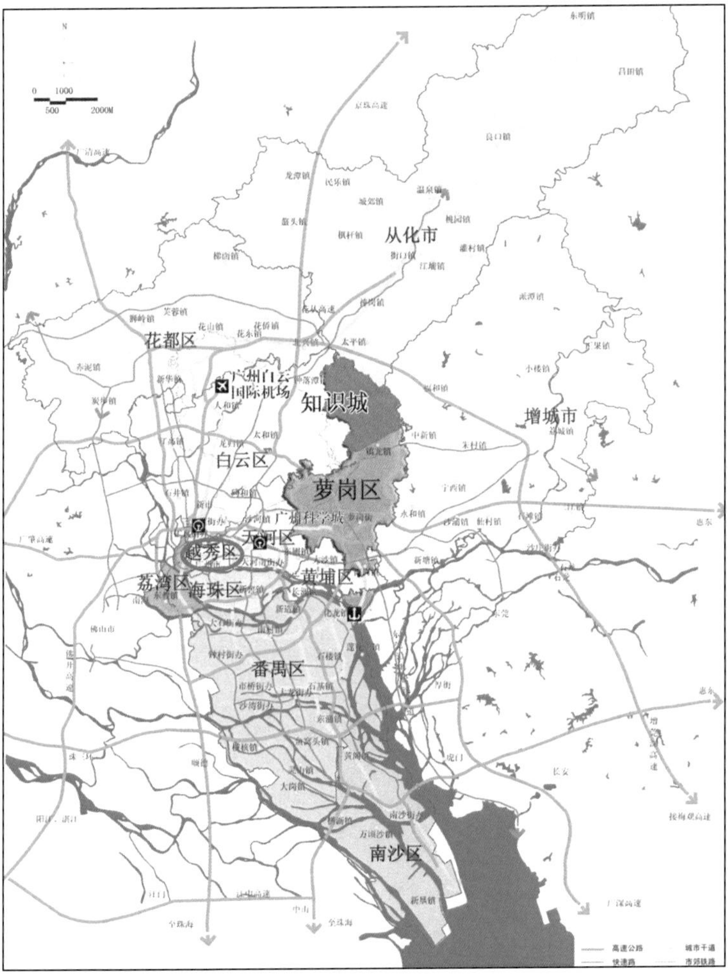

SOURCE: Sino-Singapore Guangzhou Knowledge City.
NOTE: Knowledge City is the red area, the rest of Luogang District is the orange area, and central Guangzhou City is the area within the oval at the center of the map.
RAND *MG1240-3.3*

live in a desirable area. For this reason and to place Knowledge City's growth on a sustainable path, we do not recommend policy measures to subsidize housing.

Actions for GDD

Although quality of life is important, the extent to which it precedes the creation of an innovation area versus the extent to which it is an outgrowth of an innovative area with highly educated workers is uncertain. Convenient commuting opportunities may precede the creation of an innovative area, whereas cultural opportunities, which need an audience to survive, may be an outgrowth. Nonetheless, there are a number of steps that GDD can take to increase quality of life as Knowledge City gets started. To ensure a high quality of life in Knowledge City, we recommend the following actions be taken in the sequence in which they are listed:

- *Improve transportation.* Ensure that light rail and metro links are completed between Knowledge City, Guangzhou City, Science City, and other parts of the Guangzhou area and that trains run frequently (see Figure 3.4).

Figure 3.4
Example of Transportation Infrastructure That Could Link Guangzhou City and Knowledge City

SOURCE: Sino-Singapore Guangzhou Knowledge City.
RAND *MG1240-3.4*

- *Improve schools.* Set up the highest-quality primary and secondary schools in Guangzhou in Knowledge City—schools that parents throughout Guangzhou will want their children to attend—by recruiting high-quality teachers and principals. We recognize that GDD has already taken this step for Science City by establishing the Guangzhou No. 2 Middle School there. However, a similar step needs to be taken in Knowledge City so that parents will benefit from having high-quality schools in that part of Luogang District as well.
- *Improve shopping opportunities and amenities.* Attract a destination mall near a light rail or metro station that includes comfortable venues where people can gather, such as restaurants, karaoke bars, and places for playing board games. Such gathering places can foster and reinforce networks. Businesses should also be located close to transit stations, enabling workers to commute easily and then gather at nearby stores, restaurants, and entertainment venues after work to share ideas.
- *Make Knowledge City an exciting destination.* Prospective investors in malls and other shopping amenities will be reluctant to risk investment without the prospect of an adequate return. Hence, Knowledge City faces a potential problem of investors waiting for customers and customers waiting for investors. GDD may reduce the perception of risk and enhance local amenities by making Knowledge City, itself, a destination that people will want to visit or live in. Monthly sponsored events, such as food fairs or cultural festivals, can raise the profile of Knowledge City among Guangzhou's sophisticated young residents, earning Knowledge City a reputation as the place to be and reducing the barriers to prospective investment in local businesses.

Talented Individuals, Returnees, Labor Flexibility, and Networks

In each of our three case studies, one common factor that helped start and sustain the cluster was an existing pool of highly skilled labor.

Many other case studies of high-technology clusters have similarly documented the importance of highly skilled labor, improvements in education, and investment in public research (see, e.g., Arora, Gambardella, and Torrisi, 2004; Athreye, 2004; Saxenian, 2004; Yamamura, Sonobe, and Otsuka, 2003).

However, the main source of highly skilled labor was different in each of our case studies. In the case of Silicon Valley, Stanford University and the University of California, Berkeley, provided a steady stream of skilled engineers and managers. In Maryland, highly skilled workers came from the National Institutes of Health and other federal laboratories and agencies, as well as associated private contractors. In the case of Israel, the technologically oriented units in the Israel Defense Forces and, later, multinational corporations, such as Intel and Microsoft, provided a pool of skilled workers.

In each of our case studies, both immigration and homegrown institutions played an important role in providing skilled labor. In Silicon Valley, immigrants constitute a large share of the science and engineering workforce and a growing proportion of entrepreneurs (Saxenian, 1999). Our case study of Israel also highlights the importance of nationals living abroad in driving cluster formation. For example, an Israeli-born engineer who worked for Intel was instrumental in bringing Intel to Haifa (Senor and Singer, 2009). Our interviews indicate that such connections between Israeli entrepreneurs and Israeli nationals in Silicon Valley and other parts of the United States were important in attracting multinational corporations. This phenomenon is not confined to Israel: Saxenian and Hsu (2001) argued that the Taiwanese-born, U.S.-educated engineers who form a link between Silicon Valley and Taiwan were to a large extent responsible for the success of Taiwan's Hsinchu cluster.

Our research indicates that a pool of highly skilled labor is critical to the formation of high-technology clusters. Excellent local universities are one but not the only potential source of such labor.

Conditions in GDD

There is no doubt that GDD has a strong base of business-oriented human capital. GDD also appears to have serial entrepreneurs (see

Figure 3.5). Almost half of all company founders said that they had previously founded at least one company, while more than 40 percent of company founders were first-time entrepreneurs. Given the presence of a cohort of serial entrepreneurs, this history suggests that some of these new entrepreneurs are likely to become serial entrepreneurs themselves.

The Guangzhou area is making strong efforts to develop human capital. Guangzhou has a very high level of school enrollment. Our interviews and surveys suggested that employers find it easy to recruit and train technical workers. In many cases, these workers come from local universities, including Sun Yat-Sen University and South China University of Technology. The GDD-RAND Knowledge City Project Survey indicated that in approximately one-quarter of firms, the majority of employees have a degree from a technical secondary school or junior college.

Despite the ready supply of technicians, there appears to be a gap in the availability of the most highly skilled workers, including postdoctoral candidates from top universities and of research leaders with

Figure 3.5
Prior Experience of GDD Company Founders

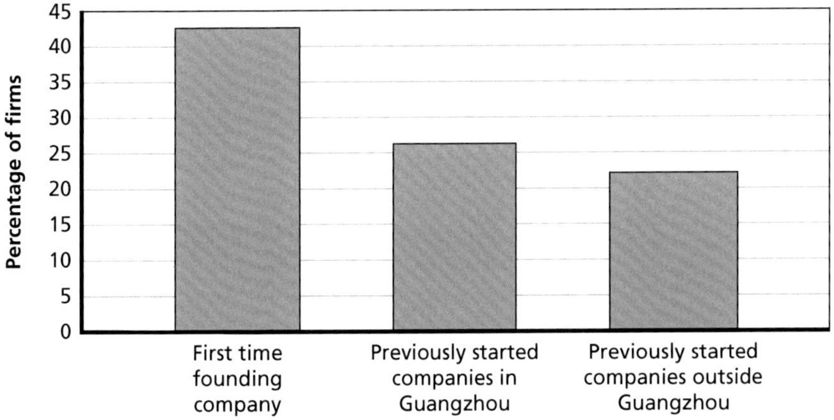

SOURCE: GDD-RAND Knowledge City Project Survey.
NOTE: Firms could give more than one answer to this question. Results are based on responses from 289 firms (out of 305 surveys received).
RAND *MG1240-3.5*

experience leading teams that are engaged in cutting-edge innovation. Our interviews and the GDD-RAND Knowledge City Project Survey found that high-technology companies in GDD have sometimes had trouble attracting the most advanced knowledge workers, such as those who have completed postdoctoral training at U.S. universities. This may be related to the quality of the area universities. Although Guangzhou has a large number of universities, some of which are quite good, as noted earlier, China's top universities are clustered elsewhere. According to the Shanghai university rankings in 2011 (Shanghai Ranking Consultancy, 2011), Beijing has the top two Chinese universities, Hangzhou has the fourth-ranked university, Shanghai has the fifth and sixth, and Tianjin has the eighth. The alternative ranking from a team at the Chinese Academy of Management Sciences that placed Sun Yat-Sen University at number 7 also placed four of the universities in the Yangtze River Delta area in its top ten, two in Beijing and two in Hubei Province ("Top 100 Chinese Universities in 2011," undated).

Aside from local universities, leading sources of skills include workers from elsewhere in China and returnees from abroad. The GDD-RAND Knowledge City Project Survey indicated that GDD companies tend not to recruit from Shanghai and Beijing, where some of the top universities are located, and interviews indicated that companies have trouble attracting educated workers from the Yangtze River Delta and Beijing. This suggests that recruiting from the top universities elsewhere in China will not fill the skills gap and that policy should focus on attracting returnees in the short term and improving local universities over the medium-to-long term. This may change as Knowledge City develops.

GDD has a number of policies aimed at attracting talented workers, and returnees in particular. These policies have had some success. GDD has attracted 2,000 returnees, and 700 companies there have been founded by returnees. One program aims to attract selected overseas returnees by offering a number of incentives from GDD, including money for starting a company, free or subsidized rent for the company, access to facilities (such as conference rooms, fax machines, and printers), and even subsidized housing and school fees. GDD has also established a "100 leading talents" policy to attract 100 highly skilled workers from across the globe. The national government has a similar "1,000 talents plan" to attract 1,000 highly skilled workers from outside China.

GDD's policies to subsidize incubator space and housing for leading talent were mentioned as advantageous but have not been sufficient to overcome all the challenges of recruiting and retaining a highly skilled workforce. These challenges are not unique to GDD: Our interviews with government officials indicated that although the Guangzhou government had a number of applicants for its grants for leading talent, few applicants met the award criteria.

Once talented individuals relocate to Knowledge City, it may be to their advantage, and to the advantage of companies in Knowledge City, if they change jobs periodically to find more rewarding work. These individuals are likely to quit jobs if a more attractive position becomes available. Such job changes can result in information spillovers—the sharing of specialized knowledge learned at one company with another.

Survey evidence indicates that there is labor mobility among high-technology firms in GDD. Top destinations for employees who leave a high-technology firm in GDD include unrelated companies in Guangzhou and competitors in Guangzhou (see Figure 3.6).

The fact that such a large fraction of employers reported competitors among the top two destinations for former employees suggests that information spillovers in GDD are common. Horizontal spillovers (movements of employees to competing companies) appear to be more common than vertical spillovers (movements of employees to suppliers or customers), although 16 percent of employers reported that suppliers or customers in Guangzhou were among the top two destinations for employees.

When employees leave, some may take proprietary company IP with them. Companies protect themselves with non-compete clauses in employment agreements. If these agreements are too strict, they may hurt the innovation potential of an area. The broader empirical evidence generally finds that areas that restrict the enforcement of non-compete clauses have higher labor mobility and higher rates of entrepreneurship.

Figure 3.6
Destinations for Former Employees

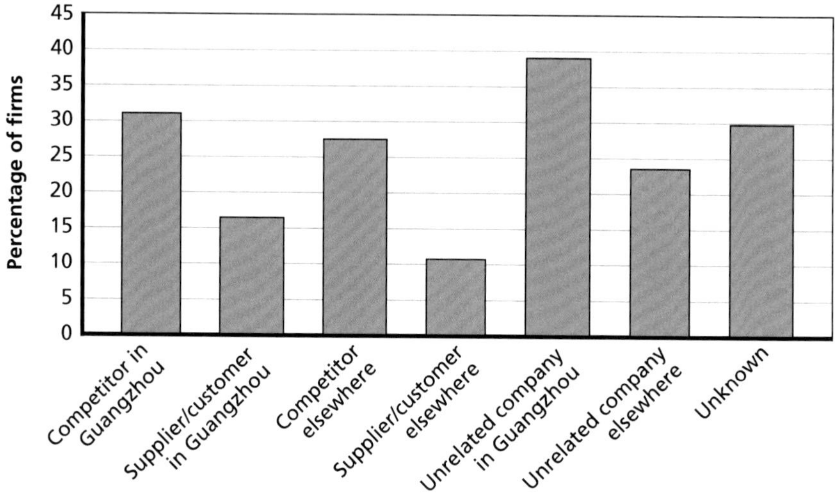

SOURCE: GDD-RAND Knowledge City Project Survey.
NOTE: Firms could give up to two answers to this question, so the total sums to more than 100 percent. Results based on responses from 284 firms (out of 305 surveys received).
RAND *MG1240-3.6*

The literature and our case studies suggest that the positive spillovers from labor mobility outweigh the losses to individual firms.

Strong IP protection through other means, such as patent protections and effective enforcement through courts, can make it possible for GDD and Knowledge City to have weaker non-compete clauses. Fortunately for GDD, labor mobility does not appear to be a concern for most employers: In the survey, more than 70 percent of respondents indicated that it would be acceptable for an employee to leave and work for a competitor; more than 85 percent reported the same for a supplier or customer. Furthermore, more than 80 percent of employers reported that they would hire employees who had previously worked for competitors, suppliers, or customers.

Labor mobility does not apply to talented individuals only. Labor flexibility, or the ability of companies to release labor that is no longer needed, is also an important factor in the development of healthy clusters. The vast majority (85 percent) of representatives from high-technology companies noted in the GDD-RAND Knowledge City Project Survey that it was difficult to lay off workers (see Figure 3.7). Many also said that it was not easy to shut down a company. Challenges in laying off workers are not unique to GDD. China's 2008 Labor Contract Law limits the hiring of short-term workers without contracts, although the enforcement of this law may vary across regions in China (Wang et al., 2009). China has also developed policies that discourage layoffs (Chan, 2012). Allowing labor flexibility and ease of business failure will aid the innovation environment by letting entrepreneurs respond more quickly to changing business conditions and to abandon innovations that do not work well. However, GDD will want to make sure that workers who lose their jobs have a good chance of finding new jobs.

Figure 3.7
Ease of Doing Business in GDD

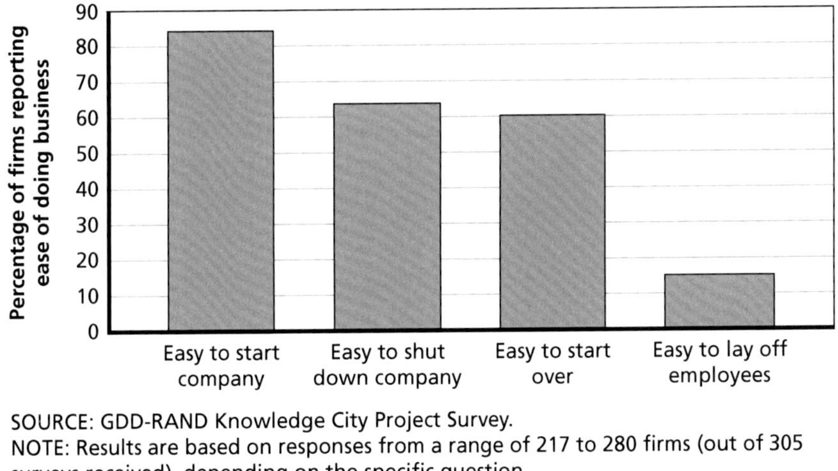

SOURCE: GDD-RAND Knowledge City Project Survey.
NOTE: Results are based on responses from a range of 217 to 280 firms (out of 305 surveys received), depending on the specific question.
RAND *MG1240-3.7*

Talented individuals and their companies can become more effective through the knowledge and professional connections that they gain from networks. Networks can enhance the flow of ideas and the abilities of individuals. Both our review of the literature and our case studies suggest that networks are a fundamental condition for a successful cluster.

Several interviewees mentioned GDD policies that have helped them build their supplier, buyer, and support networks. Approximately 20 percent of survey respondents indicated that GDD provides help in connecting to customers, while 40 percent indicated that GDD provides help in connecting to companies and investors. Advantageous policies mentioned during our interviews included arranging seminars by outside experts on such topics as patents and funding applications, providing information about other companies in the area, and offering help in finding local suppliers. Some employers were interested in additional help with networking; approximately 20 percent of survey respondents ranked help in creating networks to find new customers among the top three new policies that GDD could undertake.

Our findings about innovative areas worldwide suggest that, while networks initiated through official channels may be useful, the most important networks are likely to be created by individuals as the cluster develops. Most of the networks in the case studies we reviewed were informal; they were driven by social and business interactions among individuals rather than by alliances among firms or other institutions. Individual ties dominate institutional ties even when networks are based on relationships developed through shared formal institutions.

GDD should take advantage of opportunities to foster the growth of networks so that Knowledge City can fully benefit from the presence of talented individuals. However, given the importance of spontaneous or informal networks, it may be most valuable for GDD to focus its efforts on creating specific types of networks that may be less likely to form spontaneously. It can do so through several groups of people, including those who have received benefits from GDD or Guangzhou Municipality and through overseas Chinese professionals.

Actions for GDD

In the areas of talent, returnees, labor mobility, and network formation, we recommend that GDD undertake the following actions:

- Increase efforts to identify and attract returnees, especially those who have advanced education in technical fields, management, or finance or who have international business experience. GDD may want to consider hiring a professional recruitment firm to assist with these efforts.
- Ensure that non-compete clauses in employment contracts do not hamper labor mobility but still protect intellectual property. If GDD protects IPR strongly, firms' non-compete clauses can be weaker. The GDD-RAND Knowledge City Project Survey indicated that there is already a great deal of movement by employees among local competitors. GDD should ensure that such movement remains possible in Knowledge City.
- Enhance labor flexibility by working with national and provincial authorities to improve the legal environment for laying off workers and shutting down failing businesses.
- Work on a local and regional basis to address and smooth out the non-legal aspects of laying off workers and shutting down failing businesses. Possible solutions include training and job placement programs for dislocated workers.
- Work with the people who have received benefits from GDD to create networks. Although the most effective networks are created through informal connections among individuals who interact socially and professionally, GDD can encourage this process by creating opportunities for people who have received benefits from GDD to meet and discuss their businesses with each other. Such groups include returnees, people who have received GDD R&D grants, or people who are part of special talent programs.
- Work more actively with overseas Chinese professionals from Guangzhou. Overseas networks can foster the development of a cluster, as demonstrated by Israel and the Hsinchu cluster in Taiwan. GDD should sponsor special conferences for overseas Chinese professionals from Guangzhou, or even Guangdong Province, to introduce them to the products and innovations being developed in Knowledge City. Incentive packages should be considered to facilitate collaboration and the opening of trade channels.

- Capitalize on talent in place. Establish a Knowledge City advisory council of investors and successful entrepreneurs to meet quarterly and advise the GDD Administrative Committee on needed policy measures.

Summary of Actions for Attracting and Retaining People

The separate actions to attract and retain entrepreneurial, highly skilled people will reinforce each other. Creating an environment with quality transportation, schools, shopping, and entertainment will make Knowledge City more attractive for the people targeted in the GDD 100 Leading Talents Program, the national 1,000 Talents Plan, and other such programs. Quality shopping, dining, and entertainment options will provide the venues to spur the growth of networks that encourage innovative industries to flourish. Policies making layoffs and shutdowns smoother and less costly will lower the cost of failure and encourage entrepreneurs to try again and employees to seek new opportunities.

Actions to attract talent interact positively with actions to attract and grow high-technology companies. The presence of highly skilled people will make Knowledge City a better site for investment, and the presence of innovative companies will create employment opportunities for highly skilled workers. Attracting a research institution as an anchor should increase the level of research being conducted in the Guangzhou area and bring in a new set of researchers. Having a zone of strong IPR enforcement will enable Knowledge City to have more flexible non-compete agreements. The related movement of employees should generate valuable knowledge spillovers and lead to more innovative companies.

CHAPTER FOUR

Financing

Innovation-friendly financing forms the third pillar of an innovation system, supported by the legal, regulatory, and business support systems. Recommended actions are in two domains:

- encouraging greater availability of innovation-oriented financing
- improving other forms of financing.

This chapter briefly describes the financing challenges for high-technology firms in GDD and Knowledge City. The recommended actions are designed to make available the necessary financing on a commercial basis with appropriate regard to risk.

GDD's Assets

GDD has worked hard to create financing opportunities for innovative companies. For example, it has set up Guangzhou GET Co. Ltd., GDD's venture capital firm, and the Zhongke Baiyun Fund. The four large state-owned commercial banks have branches in GDD and extensive branch networks throughout China. The global financial center of Hong Kong is a short drive away. Guangzhou is home to many investors with substantial fortunes, a potential untapped resource for providing financing for innovative businesses.

Encouraging Greater Availability of Innovation-Oriented Financing

Our case studies suggest that financing is an important component of a successful cluster. In Silicon Valley, early inventors were financed in part by government purchases of their products and in part by individual angel investors. More organized financing, including an organized angel investor group and small business investment companies, the predecessors of venture funds, became available at the start of the transistor era in the 1950s and 1960s. In Israel, the government's Yozma program to facilitate local venture funding coincided with the rapid growth of the ICT sector.

Firm financing can be divided conceptually into several broad stages. During the basic research or technology creation phase, most funding tends to come from government agencies, universities, or similar funding sources. As the firm begins to develop the technology prior to commercialization, it faces a "valley of death." This is the period when the entrepreneur may have to rely on personal funds, or funding from friends, family, or angel investors, to survive. Once the technology is ready for commercialization, venture capital firms may step in to fund some fraction of firms. Finally, a few firms progress to the point that they receive funding from public markets through an initial public offering or can support themselves through sales (Murphy and Edwards, 2003).

Local financiers can be valuable because they can more easily find and monitor potential and ongoing investments than can financiers from elsewhere. However, it is not clear that a cluster needs its own local venture capital industry at the start. In the Israeli case, the Yozma program leveraged foreign funding by creating ten private venture capital funds, which were required to combine money from well-established Israeli institutions with that from foreign financial institutions. In Maryland, many of the venture capital firms that invested in local companies during the 1980s and 1990s were not located in the state.

Conditions in GDD

Our interviews suggest that entrepreneurs in GDD have challenges accessing early-stage financing, or financing at the point when a technology has been developed but is not yet commercialized. This will likely be the case for entrepreneurs in Knowledge City. We cannot verify whether this is so because such financing is unavailable, because the companies with promising prospects are rare, or because this is the normal state in most innovative areas. The GDD-RAND Knowledge City Project Survey shows that the vast majority of entrepreneurs used their own money as part of their initial funding and that subsequent funding, when it came from outside the firm, most often came from a bank rather than a more risk-oriented financial institution, such as an angel investor or a venture capital firm. Angel investors, in particular, appear to invest in a very small number of firms relative to the total population of high-technology firms in GDD (see Figures 4.1 and 4.2). Although this may be the case in most of China, our case studies of

Figure 4.1
Sources of Initial Funding

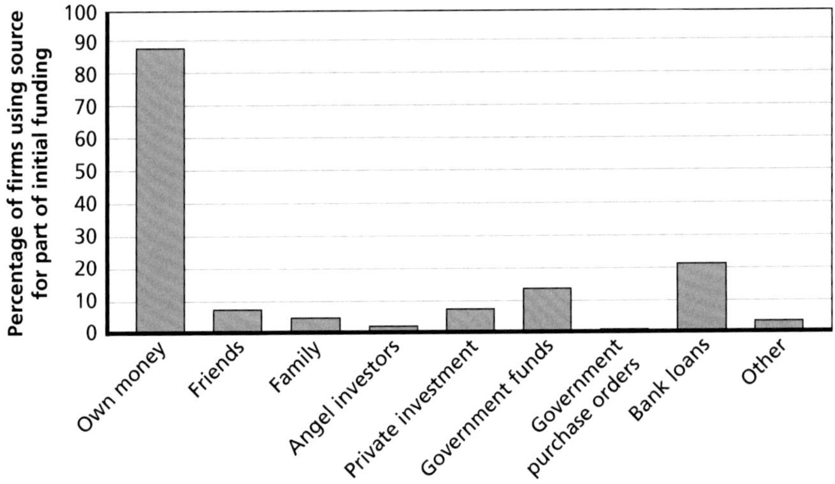

SOURCE: GDD-RAND Knowledge City Project Survey.
NOTE: Firms could give more than one answer to this question, so the total sums to more than 100 percent. Results are based on responses from 293 firms (out of 305 surveys received).
RAND *MG1240-4.1*

Figure 4.2
Sources of Subsequent Funding

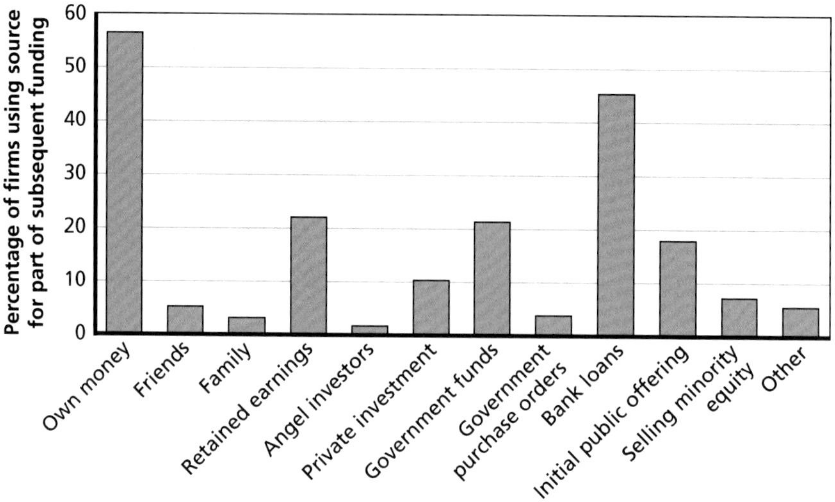

SOURCE: GDD-RAND Knowledge City Project Survey.
NOTE: Firms could give more than one answer to this question, so the total sums to more than 100 percent. Results are based on responses from 290 firms (out of 305 surveys received).
RAND *MG1240-4.2*

Silicon Valley and Maryland showed that at least some innovative areas have active angel investment networks, which GDD lacks.

GDD does provide some early-stage funding in the form of research grants for companies. However, our interviews suggest that, although such funding is viewed as helpful, there are concerns among both entrepreneurs and individuals engaged in disbursing the grants. On the one hand, some entrepreneurs indicated that the amounts provided are insufficient to ensure the development of the firm. Some entrepreneurs reported difficulty obtaining sufficient amounts of these funds, particularly when they are targeted at specific types of projects. On the other hand, individuals engaged in disbursing grants have commented that outcomes from these grants were hard to measure, and the available data did not indicate that the program was highly successful. It was not clear from either entrepreneurs or grantmakers whether an innovation would have been developed even if the grant had not been made available. It was also not clear whether the innovations that were developed generated sales commensurate with the grant support received by the company. In other words, we cannot confirm that the current policy generates a satisfactory rate of return in terms of sales. This places GDD in a similar position to many other government entities. The research evidence is mixed regarding the success of government funding in the financing of innovative firms.

The Maryland case study offers some suggestions for how GDD might facilitate such investment. In Maryland, the Maryland Technology Development Corporation (TEDCO), an agency funded primarily by the state, and the Dingman Center for Entrepreneurship at the R. H. Smith School of Business at the University of Maryland have established angel investor groups. These organizations invite investors in their networks to attend regular sessions during which selected companies pitch their ideas. The angel investors are typically successful local entrepreneurs but may also include wealthy individuals who have not had previous entrepreneurial experience.

TEDCO invites approximately 20 companies from its portfolio of firms and incubators to pitch their ideas to TEDCO officials. These officials then select the top ten to pitch to the angel group. Approximately 25 angel investors attend each meeting; companies have ten minutes to pitch their ideas, followed by a networking event. Angel investors in the Dingman network pay a small fee to be part of the group, since their membership allows them access to a constant flow of high-quality deals.

GDD could set up such a forum to provide a platform for angel investors and entrepreneurs to meet on a regular basis. GDD could also hire advisers to educate entrepreneurs on how to prepare presentations. GDD leaders would need to attend these meetings to underline their importance to potential investors. To draw on existing investment efforts, GDD could coordinate with Guangzhou GET Co. Ltd. to identify individual investors, arrange meetings, and select firms to present. Companies invited to pitch their ideas to the angel investors could be selected from among the high-technology firms in GDD.

Actions for GDD

GDD should not try to fill the financing gap on its own. The record of successful government funding of small high-technology companies is often poor. Instead, we recommend the following actions:

- *Assist in the creation of angel investor networks.* GDD can assist in the formation of these networks by mobilizing the existing entrepreneur-investor base and engaging wealthy Guangzhou investors in these networks. We strongly recommend that GDD first seek advice and assistance from people who have already created such networks. Steps in creating an angel investor network would include
  - contacts between high-level GDD staff and wealthy individuals
  - regular meetings, perhaps monthly, between investors and entrepreneurs
  - coordination between GDD and entrepreneurs before meetings to ensure that proposal presentations are professional and attractive to investors
  - facilitation of follow-up meetings between investors and entrepreneurs
  - gathering feedback from the investors to improve the operation of the network.
- *Facilitate private venture capital.* Facilitation may include recruiting foreign or domestic venture capital firms to target companies located in Knowledge City. GDD may wish to pursue Israel's Yozma model, in which GDD would invest in foreign venture capital firms that would then make the investment decisions. For example, Guangzhou GET Co. Ltd., GDD's venture capital firm, might be encouraged to invest in several foreign or private venture capital firms that target high-technology companies in Knowledge City.

Other Types of Financing

Although we have concentrated on angel investments and venture capital, those are not the only forms of capital that growing firms in Knowledge City will need. They will still need other forms of financing, including regular commercial bank loans.

Conditions in GDD

As noted earlier, firms in GDD rely on bank loans for at least some of their external financing. Approximately 25 percent of GDD-RAND Knowledge City Project Survey respondents indicated that they had received some outside funding. Among these firms, banks and bank loans were the most common source in the initial round of outside funding, followed by investment funds, including venture capital, private equity, and other investment funds (see Figure 4.3). Five percent of respondents received their first round of outside funding from angel investors.

This suggests that commercial banks fill an important financing gap in GDD. Even if this gap is filled through more angel investment and venture capital opportunities, banks will continue to play a valuable financing role.

Actions for GDD

Guangzhou is home to several commercial banks. The challenge will be access to these banks by firms in Knowledge City. Therefore, we recommend the following action:

- *Bring commercial banking to Knowledge City.* Although commercial banks should not be a primary source of innovation-oriented financing, they are an important part of the financial ecosystem that enables innovation-oriented financing from angel investors and venture capital firms to be most effective. GDD is well supplied with commercial banking branches; Knowledge City will need such a network as well. GDD should ensure that Knowledge City companies have a choice of convenient commercial bank branches.

Figure 4.3
Sources of First-Round Outside Funding

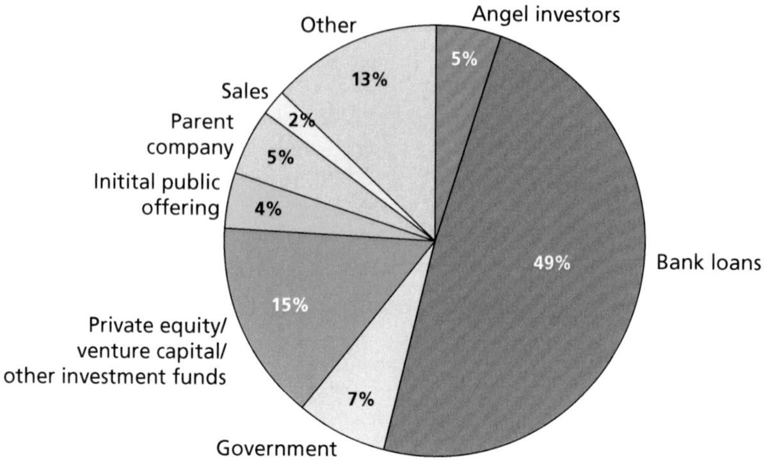

SOURCE: GDD-RAND Knowledge City Project Survey.
NOTES: Results are based on responses from 75 firms (out of 82 that reported receiving outside funding, among a total of 305 surveys received). The "Other" category includes the following reported sources: "Britain," "legal entity," "loan guarantee," "self," and "friends."
RAND *MG1240-4.3*

Summary of Actions for Financing

The actions recommended in this chapter will help fill gaps in the weakest parts of the current GDD financing system, particularly at the points of precommercial technology development (angel investors) and at the first stage of commercialization (venture capital). These actions also support other measures to attract and foster high-technology companies and to attract and retain entrepreneurial people. Better financing to attract both companies and talented individuals will trigger a virtuous circle: These talented individuals will become prime candidates to participate in angel networks once they have succeeded in business and built up capital that can be invested.

CHAPTER FIVE

Priorities, Ease of Implementation, and Sequencing

The recommendations in this report consist of a coordinated set of actions. Given limited resources, it will typically not be possible or even desirable to implement all of these initiatives at the same time. However, taking some of these actions at an early stage (for example, hiring a marketing or recruitment firm) can help facilitate other, later actions (for example, attracting returnees).

We ranked these recommended actions on three dimensions. Table 5.1 pairs each action with one of the three pillars of an innovation system: attracting companies and enabling their growth, attracting and retaining highly skilled people, and ensuring the availability of innovation-oriented financing. Within those pillars, it lists the actions in order of priority. We first ranked each action in terms of high, medium, or low priority. Low-priority actions should not be seen as unimportant; rather, they are less crucial for success than the medium- and high-priority actions. Second, we ranked each action in terms of ease of implementation (easy, medium, or difficult). In the table, we suggest potential sequencing by indicating whether each action should be performed at an early, middle, or late stage. Typically, we have suggested that actions with higher priority and those that are relatively easy to implement be taken at an early stage. For example, hiring a marketing or public relations firm is a high priority and should be fairly easy to implement, so we have suggested an early sequencing for that action. In contrast, working with national and provincial officials to improve the legal environment for laying off workers is likely to be difficult;

moreover, these challenges are faced by firms throughout China, not just in Knowledge City. Therefore, we have suggested a late sequencing.

There are exceptions to this general pattern. Although attracting an anchor institution may prove to be difficult, it should be undertaken at an early stage because such an institution can be a powerful draw for other high-technology firms and for talented individuals. In contrast, although attracting a high-quality middle and elementary school to Knowledge City is a high priority, it may prove difficult and thus should occur after several other policy actions related to attracting and retaining people are under way.

This outline of a strategy provides guidance for the next steps in the development of Knowledge City as a site of innovation. As GDD develops a full strategy based on this outline, it will want to set timelines, specific activities, and key performance indicators for each of these actions.

Table 5.1
Ranking of Priority, Ease of Implementation, and Sequencing of Actions

Action	Priority	Ease of Implementation	Sequencing
Attracting High-Technology Companies and Enabling Their Growth			
Hire a marketing or public relations company.	High	Easy	Early
Attract an anchor institution.	High	Difficult	Early
Set up an office to assist companies with all legal, administrative, and financing issues.	Medium	Medium	Early
Establish a "case officer" system for each company, eventually housing the case officers in a general assistance office.	Medium	Easy	Early
Have the office and case officer help companies apply for and receive all available tax and non-tax incentives.	Medium	Medium	Early
Ensure that non-tax incentive packages, such as grants, loan subsidies, or direct purchases by GDD on behalf of firms, can respond to the unique needs of innovative firms.	Medium	Easy	Middle

Table 5.1—Continued

Action	Priority	Ease of Implementation	Sequencing
Support initiatives by high-technology firms in Knowledge City to market their products in Hong Kong, Taiwan, Korea, and Japan.	Medium	Medium	Middle
Make Knowledge City a zone of IPR enforcement.	Medium	Difficult	Middle
Sponsor an annual international trade fair or conference.	Low	Easy	Late
Expand assistance for international patenting.	Low	Medium	Late
Attracting and Retaining People			
Establish an advisory council.	High	Easy	Early
Ensure that rail transit connections to Knowledge City are completed and that trains run frequently.	High	Difficult	Early
Increase efforts to attract returnees to Knowledge City, possibly by hiring a professional recruitment firm.	High	Medium	Early
Attract high-quality middle and elementary schools to Knowledge City.	High	Medium	Middle
Attract a destination shopping center to Knowledge City and facilitate the creation of entertainment venues.	Medium	Difficult	Middle
Facilitate the development of networks among people who have received assistance from GDD.	Medium	Medium	Middle
Facilitate cooperation with overseas Chinese professionals from Guangzhou and Guangdong Province.	Medium	Medium	Middle
Work on a local and regional basis to smooth out non-legal aspects of laying off workers and shutting down failing firms (for example, establish a job training and placement program).	Low	Medium	Late

Table 5.1—Continued

Action	Priority	Ease of Implementation	Sequencing
Work with provincial and national governments to improve the legal environment for laying off workers and shutting down failing firms.	Low	Difficult	Late
Create events to make Knowledge City an exciting destination.	Low	Easy	Late
Review and improve non-compete clauses.	Low	Difficult	Late
Ensuring the Availability of Innovation-Oriented Financing			
Assist in the creation of angel investor networks.	High	Medium	Early
Bring commercial banking to Knowledge City.	High	Easy	Middle
Facilitate private venture investment.	Medium	Medium	Middle

CHAPTER SIX

Indicators of Innovation

To ensure that Knowledge City makes steady progress toward the goal of creating and encouraging growth among high-technology firms, progress will need to be measured. Ideally, one would directly measure progress toward the desired goal of growth among high-technology firms. However, because achieving this goal may take some time, it is also important to measure progress toward creating the necessary inputs. If sufficient progress is not being made toward the goal, then measuring progress toward the creation of inputs can shed light on which elements of the innovation environment should be targeted for improvement.

In Table 6.1, we present 12 high-priority indicators that can be used to measure progress toward the goal of attracting high-technology firms and enabling their growth, as well as toward the inputs of human capital and financing. For each goal or input, there are many potential ideal measures, as well as metrics for capturing (to the extent possible) those ideal measures. Five of our suggested metrics—number of firms, output, employment, value added, and patenting activity—capture various outputs that reflect Knowledge City's growth. The remaining seven metrics capture aspects of the overall innovation environment, human capital, and financing. For example, to measure quality of life (an important factor in attracting human capital), we suggest two metrics: the population residing in Knowledge City and the share of firms that do not report that quality-of-life issues present a challenge in recruiting or retaining workers.

For each suggested metric, we identify potential methods for collecting the required data. Many of these metrics require surveys of businesses and institutions, although a few may be drawn from currently collected statistical data. We also discuss each metric's strengths and weaknesses. For the above example, the population residing in Knowledge City can be drawn from administrative data, making it easy to collect, but this metric does not reflect the extent to which firms find it difficult to recruit or retain workers with the desired characteristics. In contrast, measuring the share of firms that do not report that quality-of-life issues present a challenge in recruiting or retaining workers would require a firm survey, and the quality of the responses would depend on the extent to which the survey respondent is aware of worker concerns; however, this metric would provide information about the local workforce that is relevant to firms.

We have developed these indicators specifically for Knowledge City to reflect the various aspects of the innovation environment depicted in Figure 1.3 in Chapter One, as well as many of the specific actions described earlier. However, we have also taken care to ensure that many of our metrics reflect indicators used by other organizations. Whenever relevant, we point out similar metrics that are found in the *Silicon Valley Index* ("SV Index" in the table), published by Joint Venture and the Silicon Valley Community Foundation, or the statistical indicator database maintained by the OECD.

These indicators will provide a solid start for GDD in its efforts to monitor its inputs into the development of Knowledge City and the outcomes that will result. In the companion report, *Creating an Innovation System for Knowledge City* (Nataraj et al., 2012), we include a more complete list of 54 indicators. GDD officials can draw from these, either now or as Knowledge City develops. As GDD pursues its goals, and as Knowledge City grows, these indicators may need to be revised. Therefore, GDD should continue to scan other innovative areas for how they monitor their progress. It should also engage with international research institutes to identify best practices and new developments in measuring innovation and clusters.

Table 6.1
Indicators of Innovation

Ideal Measure	Suggested Metric	How to Measure	Strengths and Weaknesses of Suggested Metrics	Similar Metric Used in Silicon Valley or OECD Index (if applicable)
Goal: Attract High-Technology Companies and Enable Their Growth				
Action: Attract Anchor Institutions and Market Knowledge City				
Attractiveness of Knowledge City to companies	Number of firms in high-technology sectors	Administrative data	Strength: Relatively easy to measure Weakness: Does not reflect ability of companies to survive or grow	SV Index (total number of establishments); OECD (number of enterprises by sector)
Growth of companies	Output (level and growth), by sector	Firm survey	Strength: Measures commercial potential Weakness: May not reflect innovative or high-value-added products or services	OECD
	Employment (level and growth), by sector	Firm survey	Strength: Measure of local employment Weakness: May not reflect employment in innovative firms	SV Index (employment, number of jobs, unemployment rate, employment by industry, non-employer firm growth); OECD (various measures of employment) Note: Non-employers are generally self-employed people.

Table 6.1—Continued

Ideal Measure	Suggested Metric	How to Measure	Strengths and Weaknesses of Suggested Metrics	Similar Metric Used in Silicon Valley or OECD Index (if applicable)
Value-added total factor productivity	Value added per employee	Firm survey	Strengths: Reflects high-value-added products or services; relatively simple to measure Weakness: Does not take into account the productivity of other factors (capital, materials)	SV Index (value added per employee); OECD (labor productivity, total factor productivity)
Number of innovations with strong potential for commercialization	Number of patents registered in the United States	Collect administrative data from U.S. Patent and Trademark Office or conduct firm survey	Strengths: Relatively easy to collect administrative data; data generally identify inventor's city; captures quality of innovation to some extent Weaknesses: May be more difficult to identify patents associated with Knowledge City specifically; may not capture potential for commercialization	SV Index (patent registrations, patent registrations by technology area); OECD
Action: Improve the Overall Innovation Environment				
Strength of IPR enforcement	Firms' perceptions about IPR enforcement	Firm survey	Strength: Captures key measure of business climate that could help differentiate Knowledge City from other regions in China Weaknesses: May be difficult to quantify; does not reflect perceptions of firms that choose not to locate in Knowledge City because of IPR challenges	

Table 6.1—Continued

Ideal Measure	Suggested Metric	How to Measure	Strengths and Weaknesses of Suggested Metrics	Similar Metric Used in Silicon Valley or OECD Index (if applicable)
Amount of "creative destruction": opening of new firms and closing of firms that are competing poorly	Number of startup firms in high-technology industries	Firm survey or administrative data	Strength: Relatively easy to measure Weakness: Does not measure quality of firms or potential for innovation	SV Index (number of new establishments); OECD (enterprise birth rate)
		Goal: Attract and Retain People		
		Action: Improve Quality of Life		
Overall quality of life	Share of firms that do not report that quality-of-life issues present a challenge in recruiting or retaining workers	Firm survey	Strength: Reflects firms' perceptions about workers who may not choose to locate in Knowledge City because of quality-of-life issues Weakness: Firms may not accurately perceive workers' concerns	
	Population residing in Knowledge City	Administrative data	Strength: Relatively easy to measure Weakness: May not capture the population involved with innovative activities or the ability of firms to recruit or retain workers with the desired characteristics	SV Index (population, migration)

Table 6.1—Continued

Ideal Measure	Suggested Metric	How to Measure	Strengths and Weaknesses of Suggested Metrics	Similar Metric Used in Silicon Valley or OECD Index (if applicable)
		Action: Attract Talented Individuals and Returnees		
Number of talented individuals	Number of Knowledge City employees with bachelor's or advanced degrees in science, technology, engineering, mathematics, management, or finance	Firm survey	Strength: Reflects local workforce Weakness: May not reflect quality of workers or whether workers' skills match firms' needs	SV Index (educational attainment of population); OECD (tertiary educational attainment)
		Goal: Finance		
		Action: Ensure the Availability of Innovation-Oriented Finance		
Ability of firms to find sufficient early-stage financing	Amount of private venture financing (or share of total investment)	Firm survey	Strength: Can be measured midway through the firm's life cycle Weakness: May not reflect whether firms can find sufficient funding	SV Index (amount of venture capital, venture capital by industry, share of venture capital in total U.S. venture capital)
	Amount of angel investment (or share of total investment)	Firm survey	Strengths: Reflects a missing component in GDD funding; can be measured early in the firm's life cycle Weakness: May not reflect whether firms can find sufficient funding	

APPENDIX

Contents of *Creating an Innovation System for Knowledge City*

This report is a companion volume to another RAND report, *Creating an Innovation System for Knowledge City* (TR-1293-GDD), a compilation of results from the interim analyses conducted for this project and supporting evidence for the conclusions presented in this report. That volume is available at http://www.rand.org/pubs/technical_reports/TR1293.html. The contents of that volume are as follows:

Nontax Incentives
- Lessons Learned from the Case Studies and the Literature
- Conditions in GDD
- How GDD Entrepreneurs View Nontax Incentive Policies
- Gap Analysis

Intellectual Property Rights
- Lessons Learned from the Case Studies and the Literature
- Conditions in GDD
- Gap Analysis

CHAPTER NINE
Lessons Learned from the Case Studies: Other Issues

Human Capital
- Lessons Learned from the Case Studies and the Literature
- Conditions in GDD
- Gap Analysis

Infrastructure and Business Climate
- Lessons Learned from the Case Studies and the Literature
- Conditions in GDD
- Gap Analysis

Networks
- Lessons Learned from the Case Studies and the Literature
- Conditions in GDD
- Gap Analysis

Quality of Life
- Lessons Learned from the Case Studies and the Literature
- Conditions in GDD
- Gap Analysis

Marketing Knowledge City
- Lessons Learned from the Case Studies and the Literature
- Conditions in GDD
- Gap Analysis

CHAPTER TEN
Conclusions and Summary

References

Aghion, Philippe, Mathias Dewatripont, Luosha Du, Ann Harrison, and Patrick Legros, "Industrial Policy and Competition," Growth and Sustainability Policies for Europe, Working Paper 17, June 2011.

Arora, Ashish, Alfonso Gambardella, and Salvatore Torrisi, "In the Footsteps of Silicon Valley? Indian and Irish Software in the International Division of Labor," in Timothy Bresnahan and Alfonso Gambardella, eds., *Building High-Tech Clusters: Silicon Valley and Beyond*, Cambridge, UK: Cambridge University Press, 2004, pp. 78–120.

Athreye, Suma, "Agglomeration and Growth: A Study of the Cambridge High-Tech Cluster," in Timothy Bresnahan and Alfonso Gambardella, eds., *Building High-Tech Clusters: Silicon Valley and Beyond*, Cambridge, UK: Cambridge University Press, 2004, pp. 121–159.

Chan, Kay-Wah, "The Global Financial Crisis and Labor Law in China," *The Chinese Economy*, Vol. 45, No. 3, May–June 2012, pp. 24–41.

China State Intellectual Property Office, *National Patent Development Strategy (2011–2020)*, Beijing, 2010.

Cyranoski, David, "China's Patent Push: Asia Defies Patent-Filing Downturn as Global Economy Slips," *Nature*, February 15, 2010. As of August 9, 2012: http://www.nature.com/news/2010/100215/full/news.2010.72.html

Dahl, Michael S., and Olav Sorenson, "The Migration of Technical Workers," *Journal of Urban Economics*, Vol. 67, No. 1, January 2010, pp. 33–45.

European Commission, *Innovation Policy in a Knowledge-Based Economy*, Luxembourg, EUR 17023, June 2000. As of August 22, 2012: ftp://ftp.cordis.europa.eu/pub/innovation-policy/studies/studies_knowledge_based_economy.pdf

Greenstone, Michael, Richard Hornbeck, and Enrico Moretti, "Identifying Agglomeration Spillovers: Evidence from Winners and Losers of Large Plant Openings," *Journal of Political Economy*, Vol. 118, No. 3, June 2010, pp. 536–598.

Hospers, Gert-Jan, Frederic Sautet, and Pierre Desrochers, "Silicon Somewhere: Is There a Need for Cluster Policy?" in Charlie Karlsson, ed., *Handbook of Research on Innovation and Clusters: Cases and Policies*, Cheltenham, UK: Edward Elgar, 2008, pp. 430–446.

Keupp, Marcus Matthias, Angela Beckenbauer, and Oliver Grossman, "How Managers Protect Intellectual Property Rights in China Using De Facto Strategies," *R&D Management*, Vol. 39, No. 2, March 2009, pp. 211–224.

Kolko, Jed, David Neumark, and Marisol Cuellar Mejia, *Business Climate Rankings and the California Economy*, San Francisco, Calif.: Public Policy Institute of California, 2011. As of August 22, 2012:
http://www.ppic.org/main/publication.asp?i=867

Kroll, Henning, *Indicator-Based Reporting on the Chinese Innovation System 2010: The Regional Dimension of Science and Innovation in China*, Karlsruhe, Germany: Fraunhofer ISI, Discussion Papers in Innovation Systems and Policy Analysis, No. 25, 2010.

Kroll, Henning, and Ulrike Tagscherer, *Chinese Regional Innovation Systems in Times of Crisis: The Case of Guangdong*, Karlsruhe, Germany: Fraunhofer ISI, Discussion Papers in Innovation Systems and Policy Analysis, No. 19, 2009.

Liang, Mark, "Chinese Patent Quality: Running the Numbers and Possible Remedies," *John Marshall Review of Intellectual Property Law*, Vol. 11, No. 3, 2012, pp. 478–522.

Lin, Justin Yifu, "From Flying Geese to Leading Dragons: New Opportunities and Strategies for Structural Transformation in Developing Countries," Helsinki, Finland: United Nations University World Institute for Development Economics Research, WIDER Annual Lecture 15, 2011. As of August 27, 2012:
http://www.wider.unu.edu/publications/annual-lectures/en_GB/AL15/

Lin, Justin, and Célestin Monga, "Growth Identification and Facilitation: The Role of the State in the Dynamics of Structure Change," *Development Policy Review*, Vol. 29, No. 3, May 2011, pp. 264–290.

Liu, Feng-Chao, Denis Fred Simon, Yu-Tao Sun, and Cong Cao, "China's Innovation Policies: Evolution, Institutional Structure, and Trajectory," *Research Policy*, Vol. 40, No. 7, September 2011, pp. 917–931.

Lundvall, Bengt-Åke, Björn Johnson, Esben Sloth Andersen, and Bent Dalum, "National Systems of Production, Innovation and Competence Building," *Research Policy*, Vol. 31, No. 2, February 2002, pp. 213–231.

Murphy, L. M., and P. L. Edwards, *Bridging the Valley of Death: Transitioning from Public to Private Sector Financing*, Golden, Colo.: National Renewable Energy Laboratory, NREL/MP-720-34036, May 2003.

Nataraj, Shanthi, Howard J. Shatz, Keith Crane, Steven W. Popper, and Xiao Wang, *Creating an Innovation System for Knowledge City*, Santa Monica, Calif.: RAND Corporation, TR-1293-GDD, 2012. As of September 2012: http://www.rand.org/pubs/technical_reports/TR1293.html

OECD—*See* Organisation for Economic Co-operation and Development.

Organisation for Economic Co-operation and Development, *National Innovation Systems*, Paris, 1997.

"Patents, Yes; Ideas, Maybe," *The Economist*, October 14, 2010. As of August 22, 2012:
http://www.economist.com/node/17257940

Porter, Michael E., "Competitive Advantage, Agglomeration Economies, and Regional Policy," *International Regional Science Review*, Vol. 19, Nos. 1–2, April 1996, pp. 85–94.

———, "Clusters and the New Economics of Competition," *Harvard Business Review*, Vol. 76, No. 6, November 1998, pp. 77–90.

Quan, Xiaohong, and Henry Chesbrough, "Hierarchical Segmentation of R&D Process and Intellectual Property Protection: Evidence from Multinational R&D Laboratories in China," *IEEE Transactions on Engineering Management*, Vol. 57, No. 1, 2010, pp. 9–21.

Saxenian, AnnaLee, *Silicon Valley's New Immigrant Entrepreneurs*, San Francisco, Calif.: Public Policy Institute of California, 1999. As of August 22, 2012: http://www.ppic.org/content/pubs/report/R_699ASR.pdf

———, "Taiwan's Hsinchu Region: Imitator and Partner for Silicon Valley," in Timothy Bresnahan and Alfonso Gambardella, eds., *Building High-Tech Clusters: Silicon Valley and Beyond*, Cambridge, UK: Cambridge University Press, 2004, pp. 190–228.

Saxenian, AnnaLee, and Jinn-Yuh Hsu, "The Silicon Valley–Hsinchu Connection: Technical Communities and Industrial Upgrading," *Industrial and Corporate Change*, Vol. 10, No. 4, December 2001, pp. 893–920.

Schwaag Serger, Sylvia, and Magnus Breidne, "China's Fifteen-Year Plan for Scientific and Technological Development: A Critical Assessment," *Asia Policy*, Vol. 4, 2007, pp. 135–164.

Senor, Dan, and Saul Singer, *Start-Up Nation: The Story of Israel's Economic Miracle*, New York: Hachette, 2009.

Shanghai Ranking Consultancy, *Academic Ranking of World Universities 2011*, Shanghai: Center for World Class Universities and Institute of Higher Education of Shanghai Jiao Tong University, 2011.

"Top 100 Chinese Universities in 2011," china.org.cn, undated. As of August 22, 2012:
http://china.org.cn/top10/top_universities_of_China/2011-06/15/content_22787659.htm

Wang, Haiyan, Richard P. Applebaum, Francesca Degiuli, and Nelson Lichtenstein, "China's New Labour Contract Law: Is China Moving Towards Increased Power for Workers?" *Third World Quarterly*, Vol. 30, No. 3, 2009, pp. 485–501.

White, Steven, Jian Gao, and Wei Zhang, "Financing New Ventures in China: System Antecedents and Institutionalization," *Research Policy*, Vol. 34, No. 6, August 2005, pp. 894–913.

World Intellectual Property Organization, "Patent Grants by Country of Origin and Patent Office (1995–2009)," WIPO Statistics Database, Geneva, Switzerland, last updated January 2011.

Yamamura, Eiji, Tetsushi Sonobe, and Keijiro Otsuka, "Human Capital, Cluster Formation, and International Relocation: The Case of the Garment Industry in Japan, 1968–98," *Journal of Economic Geography*, Vol. 3, No. 1, January 2003, pp. 37–56.

Yang, Deli, "The Development of Intellectual Property in China," *World Patent Information*, Vol. 25, No. 2, June 2003, pp. 131–142.